# *Renewables Are Ready*

## People Creating Renewable Energy Solutions

## THE REAL GOODS INDEPENDENT LIVING BOOKS

Paul Gipe, *Wind Power for Home & Business: Renewable Energy for the 1990s and Beyond*

Michael Potts, *The Independent Home: Living Well with Power from the Sun, Wind, and Water*

Gene Logsdon, *The Contrary Farmer*

Edward Harland, *Eco-Renovation: The Ecological Home Improvement Guide*

Leandre Poisson and Gretchen Vogel Poisson, *Solar Gardening: Growing Vegetables Year-Round the American Intensive Way*

*Real Goods Solar Living Sourcebook: The Complete Guide to Renewable Energy Technologies*, Eighth Edition, edited by John Schaeffer

Athena Swentzell Steen, Bill Steen, and David Bainbridge, with David Eisenberg, *The Straw Bale House*

Nancy Cole and P.J. Skerrett, the Union of Concerned Scientists, *Renewables Are Ready: People Creating Renewable Energy Solutions*

Real Goods Trading Company in Ukiah, California, was founded in 1978 to make available new tools to help people live self-sufficiently and sustainably. Through seasonal catalogs, a quarterly newspaper (*The Real Goods News*), the *Solar Living Sourcebook*, as well as retail outlets, Real Goods provides a broad range of renewable-energy and resource-efficient products for independent living.

"Knowledge is our most important product" is the Real Goods motto. To further its mission, Real Goods has joined with Chelsea Green Publishing Company to co-create and co-publish the Real Goods Independent Living Book series. The titles in this series are written by pioneering individuals who have firsthand experience in using innovative technology to live lightly on the planet. Chelsea Green books are both practical and inspirational, and they enlarge our view of what is possible as we enter the next millennium.

Ian Baldwin, Jr.
President, Chelsea Green

John Schaeffer
President, Real Goods

# Renewables Are Ready

## People Creating Renewable Energy Solutions

**NANCY COLE AND P.J. SKERRETT**

**The Union of Concerned Scientists**

with Kevin Gallagher, Gunnar Hubbard, Robin Sherman,
Kevin Weng, and Victoria Chanse

**A REAL GOODS INDEPENDENT LIVING BOOK**

CHELSEA GREEN PUBLISHING COMPANY
White River Junction, Vermont

The glossary is adapted from *The Independent Home,* by Michael Potts
(see the Resource Guide, page 227).

Designed by Jill Shaffer

**Library of Congress Cataloging-in-Publication Data**
Cole, Nancy, 1950–
    Renewables are ready—people creating renewable energy solutions /
Nancy Cole and P.J. Skerrett with Kevin Gallagher . . . [et al.].
      p.    cm.—(A real goods independent living book)
   Includes bibliographical references and index.
   ISBN 0–930031–73–3
   1. Renewable energy sources.  I. Skerrett, P.J. (Patrick J.),1953–  .
II. Title.  III. Series.
TJ808.C65   1995
333.79'4—dc20                        95-6057

Chelsea Green Publishing Company
P.O. Box 428
White River Junction, Vermont 05001

# Contents

# *Acknowledgments*

**R**enewables Are Ready originated as an idea of the Public Outreach staff at the Union of Concerned Scientists, who wanted to help a few communities develop renewable energy projects. However, delving into the task of selecting a few good project sites convinced us that hundreds—if not thousands—of local renewable projects were already under way. We clearly didn't need just a few more renewable energy projects. What we *really* needed was something that would pull together all the good work already taking place, so that people in Iowa were informed about and empowered by renewable energy activism in Vermont. We needed to inspire and motivate other people to take up the work of changing this country's energy policies and practices. And we needed a strategic framework that would maximize the effectiveness of any local action—so renewable energy advocates could make good choices about which projects to pursue.

And so, *Renewables Are Ready* was born. Soon after its inception, science writer Pat Skerrett joined our team as the book's co-author. Pat took on the gargantuan task of transforming the reams of facts and ideas into the stories of real people and real communities that are making a difference. This book is the product of the hard work and dedication of many people, several of whom I would like to acknowledge and thank here:

Kevin Gallagher, UCS intern extraordinaire, who helped develop the book's concept, scouted out many of the book's best stories, collected many of the great photos, and always held fast to the belief that grassroots energy activism would help change the world.

Gunnar Hubbard, one of UCS's most accomplished interns, who also helped from the beginning with the book's concept and research, and whose expertise made the NOAH project (see Chapter 7) the great success that it was.

Robin Sherman, Public Outreach staffer, who was always ready to help out with the seemingly never-ending tasks of research and fact-checking.

Kevin Weng, the UCS intern who single-handedly tackled the Resource Guide, tracking down references, reports, and organizations with great determination.

Vikki Chanse, the UCS intern who carried the bulk of the job of fact-checking the stories and proposing necessary adjustments to a sometimes snarly author.

Alexa Majors and Pushpam Jain, two UCS volunteers who identified and pursued information on many good stories.

Michael Brower, Michael Tennis, Donald Aitken, and Eric Denzler, UCS's energy analysts and my colleagues, who reviewed the book for technical accuracy and whose vision for a renewable energy future has been and remains a constant source of inspiration and pride.

Jan Wager and Warren Leon, UCS colleagues whose supervision, assistance, encouragement, and editing skills made this book possible.

Ellen Stein, who came through at the bitter end when hundreds of little facts had to be confirmed and those missing photos just had to be found.

All the people whose stories are told in this book. They are the real heroes of the book, and, ultimately, of our times. Their commitment and good work, often in the face of significant obstacles, will always inspire and motivate me and, I expect, others to reach for and achieve the "impossible dream."

And finally, on a very personal note, I dedicate this book to my daughter, Amelia Blythe Brady-Cole, who was born right in the middle of the book's creation. May her energy future be bright with renewables!

Nancy Cole

# Introduction

We at the Union of Concerned Scientists believe that technologies for harvesting renewable energy— including photovoltaics, solar heating devices, wind and hydroelectric turbines, and biomass generators—are already affordable, effective, and reliable enough to be used on a huge scale in the United States. The proof can be seen in the communities described in this book, and the stories of real people who made a difference. In the following chapters we highlight the many ways in which individuals and groups across the country have begun to implement ecological, renewable solutions to our energy crisis, and we offer advice for those who wish to increase public support for renewables in their own towns and cities. Before the end of the decade, renewable energy will have been carried by community activists and conscientious consumers from the realm of fantasy to an exciting and profitable reality.

But why should we care about grassroots, local-level renewable energy ventures? Aren't the regional electric utilities and public utility commissions where the action is, in the energy field? And what about state and federal legislation, judicial processes, and administrative edicts? Well, while each of these arenas is important for renewable energy, none of them eliminates the need for good old-fashioned activist work in neighborhoods and at the community level.

It has always been committed individuals, organized together with their friends and colleagues, who have stimulated great social movements in this country. And this is what we need now—a mass movement in support of renewable energy. While over the past thirty years many creative individuals and energetic groups of people have implemented path-breaking renewable energy projects, their efforts have gone almost entirely unnoticed outside their home communities. Until now, these local efforts have not been systematically documented or described, nor have they been well publicized and strategically promoted. And since the success stories from far-flung towns, from Maine to Hawaii, have never before been gathered together, their collective impact and influence have been minimized.

And yet, in the 1990s, energy is one of the most critical areas for grassroots activism. If we as Americans are ever to reach a sustainable way of living, so that our children and their children can inherit a livable world, then we must begin today to curtail our energy use, and meet the demand for this reduced energy consumption with the judicious development and use of renewable resources. And we must think in global terms: sustainability for the developing world will only be achieved if the industrialized countries transform their own energy use and provide sensible models for energy development that are not energy-intensive, exploitative, or wasteful.

This book takes on the task of collecting and relating stories of American communities where this challenge of transforming energy use has been taken to heart. It profiles individuals who have courageously—and successfully—advanced renewable energy in their communities, and gives readers information about how to contact them. The stories focus on commercialization strategies that are amenable to and appropriate for local action. All of the ideas presented in this book can be implemented in a city, town, or neighborhood by individual or group action.

The first chapter is an introduction to the topic of renewable energy, discussing the common barriers to greater renewable energy use and the current status of various applications. It also notes several key commercialization strategies for advancing renewable energy which will be effective at the local level, and highlights the efforts that are most needed in the coming years. The core section of the book follows, consisting of five chapters with descriptions of actual renewable energy projects that have been implemented in communities across the country. Each chapter begins with a case study of a particularly notable project, followed with briefer accounts of other projects with similar goals. Each of these chapters also includes a profile of an especially interesting and inspiring activist.

The book concludes with information designed to help individuals or communities carry out a project similar to the ones described in the five core chapters. Chapter 7, the "how-to" chapter on making renewables a reality in local communities, offers practical, step-by-step suggestions for organizing and carrying out a project. Included are discussions of how to select a project, how to assess the physical and human resources in the local area, how to pull together a coalition, how to develop a manageable project plan, and how to publicize your efforts. Following this chapter are several appendices designed to assist readers who are ready to take action.

- Appendix A explains how solar technologies, wind power, biomass, and small-scale hydro projects work. If you find yourself frustrated reading a story you don't understand, just flip to the section explaining PV systems or heat-pump operations. If, on the other hand, you already know which renewable energy resources are available in your community, you might focus first on the technology to sharpen your appreciation of the technology-specific examples. Finally, the "Advantages and Disadvantages" section of each technology explanation should be helpful in evaluating project potential and effectively countering arguments against your project.

- Appendix B outlines reasons why renewable energy is so important to this country's—and the world's—future, discussing the impacts and hidden costs of fossil fuels. This appendix puts renewable energy in a wider environmental, social, and political context and should be particularly helpful in crystallizing compelling arguments in favor of your project. You might also find that this appendix serves to reinforce once again why renewable energy activism is so vital and why you are doing the energy work you do.

- Appendix C is a resource guide that provides information on government agencies, trade associations, public interest groups, funding sources, local resources, and publications that can be of help. One advantage of all the good energy work currently underway is that there are plenty of resources available—you don't need to reinvent the wheel! Further, state, regional, and national organizations may be excellent resources for information on renewables. In fact, you may want to prepare for your renewable energy project by conducting your own survey of state and local resources, so you have a more complete sense of what's available for research and support of your project.

Renewables are ready! And many citizens across the country have already begun to put them into use in a myriad of ways. We hope that their inspiring stories and the strategic project suggestions in this book will motivate you to take effective action in your own community, and that your actions will be mirrored by activists in other communities as we work together to create the necessary demand for renewable energy services and technologies. Citizen action is the lifeblood of a democracy. This book is designed to get that "blood" flowing on renewable energy action, by highlighting and encouraging concrete steps that will bring renewable energy projects from the realm of "Buck Rogers" fantasy into reality.

# *Renewables Are Ready*

## People Creating Renewable
## Energy Solutions

# CHAPTER ONE

# U.S. Energy Use at the Crossroads

**M**ost Americans understand the environmental and economic problems caused by excessive reliance on fossil fuels. They express enthusiasm for renewable energy and support policies to encourage its use. But, unfortunately, many Americans do not believe that renewable energy can be deployed on a large enough scale to displace significant quantities of oil, coal, or natural gas.

Yet renewables truly are ready! Renewable energy sources that derive their energy from sunlight, wind, oceans and rivers, and plants could provide enough energy to meet a large portion of our present and future needs. During the 1980s, renewables were ignored by the public and dismissed by the federal government; nevertheless, work on the technologies moved steadily forward. The reliability and efficiency of renewable energy equipment improved, and the cost of installing, maintaining, and running it declined. In the case of wind turbines, for example, more advanced designs, better choice of materials, and careful siting have cut the cost of generating electricity from wind to one-fifth of what it was in the early 1980s—from 25¢ per kilowatt-hour to around 5¢ per kilowatt-hour. In some locations, a utility company can now build a wind power facility that will produce electricity at a cost comparable to a new fossil-fuel–powered plant.

When the Union of Concerned Scientists and three other national energy and environmental organizations analyzed the U.S. energy supply in 1991, we demonstrated that, if coupled with strong measures to use energy more efficiently, renewable energy sources could provide more than half of the U.S. energy supply by the year 2030, at a net savings to consumers of some $2.3 trillion. Fossil fuel use would not be eliminated completely, but, with well-considered policies, oil consumption could be reduced to just one-third of current levels by 2030.

*If we don't change direction, we're likely to end up where we're headed.*
— LAO-TSE

Unfortunately, renewable energy must play a daunting game of "catch up." Because renewables are starting from a small base (solar and wind technologies currently provide much less than 1 percent of U.S. energy needs), and because their costs are still somewhat higher than fossil fuels, their near-term potential is limited unless these circumstances change dramatically. A coordinated, nationwide plan is needed to transform all this renewable energy potential into U.S. energy policy and practice.

## Barriers to Greater Renewable Energy Use

Renewable energy potential is enormous, yet all renewable energy sources combined supply only about 8 percent of our nation's energy. Why this sorry state of affairs? The main reason is that several obstacles stand in the way of renewable energy expansion. Contrary to popular perception, these barriers are *not* mainly technological; many of them are political and perceptual, which means that force of will, leadership, and experience can make a big difference. We can, in fact, change the laws, regulations, incentives, public attitudes, and other factors that make up the energy market. Until this is done, many barriers to greater use of renewables remain.

**Long-range vision.** One of the most disheartening obstacles is a lack of long-range vision on energy policy and strategy. This failure of vision and leadership is found at all levels of government, industry, and other areas of public life. During the modern heyday of renewable energy under the Carter administration in the late 1970s, the Carters—like thousands of other American families—put solar panels on the roof of their home. Yet today the renewable energy industry is struggling to recapture the interest of the public and the government. Complete disinterest in renewable energy was evident at the highest levels of government under the Reagan administration, when renewables research and development funding was cut by 80 percent. Under the Clinton administration, support for renewables has increased, but it is not clear how permanent or useful this support will be. Policies to encourage renewable energy development can be defined, but it is difficult to see how they can be implemented without a fundamental change in the attitudes of policymakers.

**Short-term profit.** Another barrier at the policy level is the apparent willingness of energy planners, providers, and regulators to imperil the future for short-term ease and profitability. Like the many sectors of our profit-oriented society, the energy market is obsessed with the short term, focusing primarily on the most immediate, most narrowly defined costs, while paying little heed to the environmental and social costs of conventional fuels. Adverse effects of fossil fuel use such as pollution, greenhouse warming, health problems, and military expenditures are very real, but are very difficult to price. Until methods are determined to consistently and reliably factor these expenses into energy pricing, the true value of renewable energy will never be visible in the marketplace.

**Consumer demand.** As consumers, we make powerful choices every time we spend our money—whether it's on our cars, our homes, or in our business decisions. Recalcitrant state and national policymakers are often prodded to make changes by a strong consumer movement demanding products, services, and policies. Unfortunately, consumer demand for renewables has been just as sluggish as the policymakers have been. During the research for this book, UCS staffers often encountered the complaints from builders: "I would put solar hot-water heaters on the new homes I build, but no one is asking for them"; from renewable energy businesspeople: "My solar thermal company is just holding on, hoping for better economic times"; and from the press: "Americans aren't willing to give up their gas-guzzling cars in favor of ones that get 50 miles to the gallon." Sound familiar? Now of course, consumers can't "demand" goods or services they don't know exist. And consumers don't know renewable energy goods and services exist, because banks, builders, utilities, and others don't do a very good job of advertising the renewable products and services they have available. So the cycle continues, and the powerful market mechanism of consumer demand isn't activated.

**Start-up capital.** Most renewable energy technologies cost a lot up front, while providing savings down the road in the form of lower fuel costs. This creates a significant capital cost barrier that further discourages interest in renewable sources. A solar water heater, for example, may cost $2,500 to purchase and install, whereas a conventional water heater may cost no more than $800. However, solar technologies cost very little to operate, whereas the major cost associated with conventional technologies is usually fuel, which will be paid later.

**Research and development funds.** Since renewable energy sources represent largely new technology, investment in research and development (R&D) is critically important to their success. However, federal funding for renewable energy R&D declined throughout the 1980s, and still constitutes only a small percentage of total federal

One significant barrier to greater renewable energy use is inadequate federal funding for research and development of these relatively new technologies. R&D for renewables has gone up and down with the political times, making it difficult for the technologies to reach full commercialization.

Source: "Renewable Energy: A New National Commitment?" Fred Sissine, 1992; and U.S. Department of Energy.

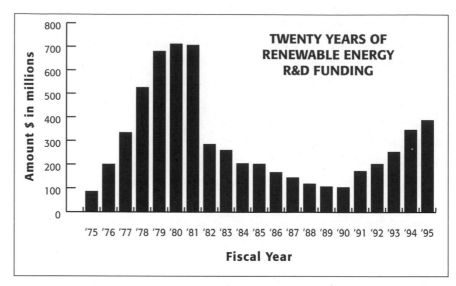

**TWENTY YEARS OF RENEWABLE ENERGY R&D FUNDING**

funding for energy supply R&D. Even when new technologies have been developed to the point of being ready for commercial testing and deployment, they have not been picked up by industry, because most renewable energy companies do not have the financial strength or technical know-how to turn basic research into reliable and marketable systems.

**Powerful interests supporting the status quo.** The very powerful fossil fuel industry stands to lose a lot of money from any changes in the current energy market. Large gas and oil companies wield considerable clout in congress and in state legislatures; combined with the smaller companies participating in the energy market, they raise fears of job losses whenever changes to their industry are discussed. Additionally, since the renewable energy industry is small and inexperienced, and tends to be locally based, its political clout is potentially diminished.

## Current Status of Renewable Energy Resources

Despite the daunting barriers described above, substantial progress in renewable energy development has been made over the past ten to fifteen years. The costs of renewable energy technologies have declined dramatically since the 1970s, and their reliability has been proven in demonstrations and commercial operation. In fact, renewable energy technologies could begin to make a significant contribution to the U.S. energy supply within the next decade or two. In some locations, wind power is already among the least expensive sources of electricity—fossil or renewable—and it is becoming less expensive every year. Passive solar design strategies can already provide substantial energy savings in new buildings at little or no extra cost. Photovoltaic systems have come down 90 percent in cost since the 1970s and have been proven reliable in a wide range of applications, from remote water-pumping stations to utility-scale power plants. Significant progress has been made in biomass and geothermal technologies as well.

At the same time, changes in attitudes and state and local regulations have created new opportunities for renewables. A few states are now requiring, through the regulatory process, that utilities engage in integrated resource planning, whereby utility planners not only compare the costs and energy contributions of energy efficiency, load shifting, and various types of generation, *but also* consider the "noncost aspects"—such as potential environmental impacts—of different resources to determine the best mix to meet future energy needs. This technique shows renewable energy in a much more favorable light, because the true total costs of electricity generated by fossil fuels becomes more visible. A handful of progressive utilities—such as the Sacramento Municipal Utility District (SMUD)—are not only aggressively promoting energy-efficiency programs (also called demand-side management, or DSM, programs), but are also pioneering renewable energy use for electricity generation. SMUD's incentive can be traced partially to the citizen-forced closing of the Rancho Seco nuclear power plant because of safety and cost considerations.

It is often easier to change the energy rules at the local level, as evidenced by the recent rash of solar access laws, upgraded building codes, city and regional energy plans, and the like. Relatively small renewable technology businesses thrive today all over the country, providing research and technical jobs in many areas. And, in true entrepreneurial spirit, nearly every community boasts the ever-resilient renewable energy entrepreneur, valiantly working to create a market for renewable products in a hostile economic environment. Many communities can count on at least one renewable-

*In some locations, wind power is already among the least expensive sources of electricity—fossil or renewable—and it is becoming less expensive every year. These turbines are spinning away at Altamont Pass, California.*

Photo credit: American Wind Energy Association.

energy–conscious builder who continues to promote passive solar design, energy-efficient construction, and perhaps even active solar systems.

In short, the overall picture is mixed, with some developments encouraging and others quite distressing. Despite the confusion, one thing is clear: this country stands at an energy crossroads, and the time has come to decide whether renewable energy will continue to play a relatively minor and auxiliary role in meeting this country's energy needs, or will assume the vital role they can and should play in America's energy future.

## The Energy Crossroads

If the United States is going to take the renewable energy road, we must make that turn right now. It would be dangerous and costly to decide twenty years from now, when fossil fuel prices are sky-high, that we must turn to solar. Industries may spring up overnight, but they do not reach maturity as quickly. And a mature renewable energy industry will be required to meet a substantial portion of our energy needs.

The most important goal of those who value renewable energy's contributions to the future must be to enhance the commercialization of renewable energy. *Commercialization* refers to the process of changing renewable energy use from a novelty to standard practice—whether we're talking about how we generate our electricity, how we build our buildings, how we run our commercial and industrial enterprises, or how we transport ourselves from place to place. With any technology, the commercialization process stimulates demand for the product, thus making it possible to capitalize on economies of scale to reduce prices. Cheaper prices, in turn, help create even greater demand. As the technology is used more frequently and in a wider range of applications, the product's reliability and reputation improves, and more potential consumers become interested.

Everyone interested in energy development, from utility regulators to low-income energy activists, can positively affect the commercialization process. Many renewable energy applications are already cost-effective, even given unfair government policies that force renewables to compete with fossil fuels on an uneven playing field. Others require help to level the playing field and create markets that will bring costs down to a competitive level. Successful commercialization requires strategic pressure on two fronts: a critical number of people capitalizing on these currently available, economically sound opportunities, *and* others working to level the playing field so that new opportunities can emerge.

## ENTREPRENEUR PROVIDES SOLAR MODEL HOMES

On his own initiative, Paul Neuffer is changing how houses get built in the greater Reno area. Neuffer Construction builds subdivisions, and the company offers four basic models—two with passive solar features and two with conventional features. Without advertising the solar features, Neuffer finds that most consumers pick the solar models. Why? Solar homes are open, spacious, and full of light—a design aesthetic that is attracting many of today's homebuyers. "No doubt about it, the solar design enhances the home's livability," says Neuffer. He is convinced that his solar models have contributed to his commercial success as well, enhancing his market by perhaps as much as 30 percent.

Since 1981, Neuffer has sold 500 passive solar homes. His best solar model is also his lowest-cost home. He explains: "Affordable passive solar homes make good sense by extending the opportunity for homeownership to those least able to qualify for mortgages." The opportunity works two ways: first, by

The passive solar design in this tract home, designed by Neuffer Construction, Reno, Nevada, delivers a more than 50-percent annual energy savings. This attractive home is the lowest-cost model offered by the builder.
Photo credit: Donald Aitken.

saving a strapped family lots of money in utility bills; and second, by making a higher percentage of the family's funds available for mortgage payments.

The passive solar designs produce tremendous energy savings. In the tract home pictured above, the passive design adds only 1 percent to the cost of construction, while delivering more than 50 percent annual energy savings.

**The role of energy efficiency.** The basic concept of renewable energy is closely linked to the belief that the world must develop sustainable, environmentally friendly energy practices. To achieve such a future, the world—and the United States in particular—must reduce overall energy use. This will be quite a trick, since poor countries desperately require greater energy use to develop, and rich countries consume energy like there's no tomorrow. Nearly all energy forecasts predict precipitous increases in energy use. Greater global use of conventional energy sources, of course, means greater health problems, pollution, economic insecurity, and perhaps even climate change.

However, energy efficiency and renewable energy can play a major role in helping to slow and eventually stop these problems. If all countries took full advantage of opportunities to improve energy efficiency, then global fossil fuel use and carbon dioxide emissions would grow slowly, if at all. And if, in addition, renewable energy sources were developed to their full potential, fossil fuel use and carbon dioxide emissions could be cut to well below today's levels, eventually approaching the 60 percent or greater reduction needed to stabilize carbon dioxide concentrations at today's levels.

Renewable energy can play the greatest role in the context of an aggressive approach to reducing energy consumption—for example, renewable energy could supply almost 53 percent of U.S. energy needs in 2030 based on a total energy consumption (with maximum efficiency) of 62 quadrillion BTUs (quads). In contrast, if energy consumption increases from its current 84 quads to 120 quads in that time period (government estimates), then the relative share of renewable energy's contribution remains small.*

Full commercialization of the renewable energy industry requires several elements:

**Availability of renewable resources.** The availability of renewable energy resources varies greatly from place to place. One of renewable energy's greatest strengths (and also a weakness, from a traditional energy point of view) is that it is a localized energy source—for example, some areas have geothermal resources, others don't. Even wind and solar resources vary considerably over short distances. For substantial electricity generation from renewable sources, the scope of the energy source must be documented over a long enough period of time to ensure its reliability. Confirmation of availability requires investments in resource exploration, trained personnel, and demonstration projects.

**Preparation of the markets.** Early scientific research demonstrates the viability of a technological idea and leads to the development of a design for a marketable device. The engineering that follows develops a practical version of the device and the associated manufacturing processes that enable it to be marketed. Early market sales lead to both manufacturing and field experience that further lower the price and increase market confidence. If the technology has proven itself, it is in a position to receive larger purchases at an ever-

---

*Alliance to Save Energy, the American Council for an Energy Efficient Economy, the Natural Resources Defense Council, and the Union of Concerned Scientists, *America's Energy Choices: Investing in a Strong Economy and a Clean Environment* (Cambridge, Massachusetts: Union of Concerned Scientists, 1991), 5.

increasing scale. Renewable energy technologies are quite varied, and each type of technology—wind power, passive solar, photovoltaics, biomass, geothermal—must go through these market stages.

In the early 1980s, California created an artificial market for wind power that drove the construction of nearly fifteen thousand machines, even though the cost was high. With experience, however, the costs came down and operators learned how to site and run the machines to maximize their performance. Now wind power is in a position to steadily expand its market share in California (with the help of a federal tax credit), and will be fully commercialized there in five to ten years.

**Support for "early adopters."** In general, people fall into several categories in terms of how they respond to change and opportunities for innovation, forming a rough bell-shaped curve: 5 to 10 percent are pioneers, those risk-takers and innovators who are often willing to purchase a product at a relatively high price; another 15 to 20 percent are early adopters, who are willing to take risks and operate somewhat ahead of the norm; and most of the population consists of followers, who will move to accept change only when they see others around them doing so, when it seems safe, acceptable, or mainstream. On the other end of the curve are the laggards and re-sisters. This change model is particularly important for renewable energy commercialization, because it suggests that both public and institutional support for those "early adopters" is critical to advancing renewable energy use into the mainstream.

**Stimulation of consumer demand.** While supporting those early adopters who are pushing renewable energy use forward, it is important to simultaneously create greater consumer demand for the products. This stimulation may be achieved through many different tactics, including high visibility of the products, direct and indirect education of potential consumers, evidence of product reliability and potential cost reductions, publicized successes of the early adopters, and a change in public attitudes to accept the need for

*The benefits of using renewable energy to supply electricity will be* most *effective if the U.S. reduces its overall energy consumption so that renewable energy sources can contribute a* relatively *larger percentage of the total.*

Sources: National Energy Strategy (NES), National Technical Information Service, 1991; and *America's Energy Choices,* Union of Concerned Scientists, et. al., 1991.

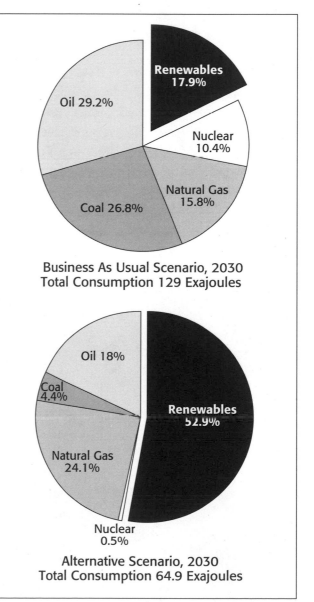

Business As Usual Scenario, 2030
Total Consumption 129 Exajoules

Alternative Scenario, 2030
Total Consumption 64.9 Exajoules

the new products (for example, convincing the public that global warming is a real threat, and thus reducing fossil fuel use is a real priority). Stimulating consumer demand for products is done every day in this country—just think about how quickly compact-disc players or video cameras have moved into the mainstream. Corporate marketers engage in consumer-demand stimulation as standard operating procedure; similar concepts need to be applied to the commercialization of renewable energy.

**Encouragement of capital investment.** The financing of renewable energy projects is a complicated matter, in part because the range of projects is so vast. What it takes to capitalize a utility-scale electricity generating project is obviously quite different from what it takes to capitalize a homeowner installing a solar water heater. Nonetheless, securing capital is a common problem faced by all manner of renewable energy projects. At minimum, renewable energy businesses must be able to see the possibility for a continuous flow of money into their projects. For example, knowing that several utility companies (including Sacramento Municipal Utility District and the New England Electric Company) want to gain institutional experience in the progressive adoption of renewable electric technologies gives the renewables industry confidence in future demand for its products. Knowing that banks offer energy-efficiency mortgages gives potential homebuyers confidence that making energy-efficient improvements in their homes can be financed over the life of the mortgage. Encouraging renewable energy capital may also mean thinking a bit differently about investments—that is, investors accepting that a return on an investment may take four or five years to be realized, or consumers accepting slightly higher electricity rates in the short term to ensure clean and affordable electricity in the long term.

**Implementation of supporting policies.** Finally, renewable energy commercialization requires changes in how we think about energy use and the policies that affect that use. Utility regulators must consider environmental damage and other indirect costs when comparing various energy sources. Legislators must redress the inequities in research and development funding and tax codes that place renewable energy resources at a disadvantage compared to conventional energy resources. Affirmative policies must be implemented to advance greater renewable energy use, such as solar access laws and high energy-efficiency standards in building codes.

Commercialization of renewable energy requires concerted and simultaneous action at the individual, community, state, regional, and national levels. Although some people believe that the federal level is the most important place for citizen pressure, in reality very few critical policy initiatives are passed at the national level without strenuous and persistent efforts at the local level. In fact, during times when federal policymakers were very conservative or hostile to changes in the status quo, state and local action were the *only* venues for effective change. Take, for example, the advances in renewable energy over the past decade, despite the flagrant hostility of the Reagan and Bush administrations.

It is certainly possible, and is in fact desirable, for community activists to take actions that will influence national and state policies—whether they involve contacting congressional representatives on the federal energy tax or intervening in public utility commission hearings to secure least-cost planning requirements for state utilities. This book, however, aims to be a catalyst for change on the local level, and focuses on those renewable energy commercialization strategies that grassroots groups can undertake in any town or neighborhood.

Five commercialization strategies are most appropriate to local action and can best advance the spread of renewable energy.

1. **Working in partnership with utilities.** Any commercialization effort must include utility companies. It is, after all, electricity generation that accounts for 44 percent of U.S. energy use; the utilities thus represent a huge market for renewable technologies. Indeed, because of the time and investment it takes to build power plants, utility companies and their regulatory commissions are often the only energy players who are engaged in long-range planning. In addition, many utility companies are potentially open to renewable strategies to diversify their energy mix. Although it may seem counterintuitive, many utilities now recognize that energy efficiency is essential to their operations: if less energy is used, then the utility can postpone and perhaps avoid altogether a huge capital investment in new power-generating capacity. Any utility that is already thinking in these terms is likely to also be experimenting with renewable energy. And one last note about utilities: municipal utilities and rural electric cooperatives, owned by their customers rather than by private stockholders, represent unique opportunities for progressive energy policies.

*This solar-powered livestock shelter and milking parlor in Maryland is a good example of a "niche market"—a specialized and relatively small market where renewable energy applications are currently cost-effective.*

Photo credit: U.S. Department of Energy.

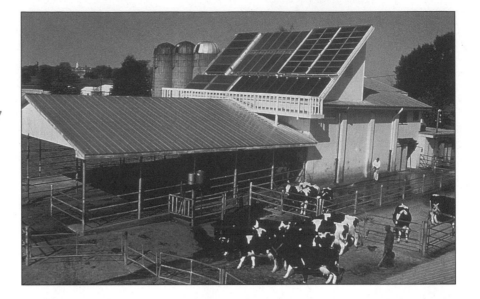

2. **Building niche markets.** *Niche markets* are those specialized and relatively small markets where renewable energy applications are currently cost-effective and therefore make economic as well as environmental sense.* By exploiting all the opportunities that currently exist, the demand for renewable energy technologies will increase, leading to cost reductions and possibly a full-blown commercial market. The classic example of a niche market is the use of photovoltaics to power a home that is a considerable distance from the nearest power line. Even though PV power is expensive per kilowatt-hour, it is still cheaper than the cost of running a transmission line all the way to the home. As unbelievable as it may seem, many cost-effective applications are not currently being pursued for lack of ac-

---

*The phrase "currently cost-effective" is used advisedly in this book because it implies that the status quo is an appropriate starting point for a discussion on the economics of energy. As Chapter 1 explains, however, renewable energy is currently at a distinct disadvantage, so the status quo is not an appropriate measure. Our current method of energy accounting does not accurately or consistently reflect the true cost of fossil fuel energy; it does not reflect the real value of renewable energy; and it does not adequately balance long-run impacts against short-term profit. So, "currently cost-effective" really means "cost-effective in the skewed and prejudiced energy accounting system that is presently in use in this country." Unfortunately, until an improved method of energy accounting is widely accepted, arguing cost-effectiveness in our current system may be necessary at times. Happily, it is true that despite the uneven playing field, many renewable energy applications are economically competitive in today's energy market.

curate information or simple technical assistance, or because of plain old inertia.

3. **Seeking creative financing.** The lack of capital to get projects started is a major barrier to greater renewable energy use. Since renewable energy applications require relatively large up-front investments, the capital barrier is particularly troublesome. Although federal tax incentives and other economic policies are important to overcoming this barrier, much can be accomplished at the local level, because to some degree, lack of knowledge about available options is part of the difficulty. In other instances, unfamiliarity with life-cycle analysis and other different ways to conceptualize projects makes potential financiers unwilling to lend or invest money. All renewable energy projects require fancy financial footwork, and there are many creative ways to get the funds together.

4. **Changing the energy rules.** Once renewable energy gains are achieved, it is critical to institutionalize them—to make sure these gains are supported by the force and authority of commonly recognized institutions. The "rules" we're talking about here range from local government codes for building and zoning to sophisticated energy planning. But changing the energy rules also means changing something far less tangible than zoning laws—public attitudes. In short, renewable energy use must change from a "freak occurrence" to "business as usual." Full commercialization means that renewable energy becomes as common and as unremarkable as any energy source. To get from here to there, the traditional rules and attitudes that affect energy use must be challenged and changed.

5. **Educating by example.** Last but not least, one thing activists can certainly do is help create the demand for greater renewable energy use. Many potential renewable energy consumers *still* are in the dark ages on renewable energy, or they think renewable energy is inadequate to handle the country's energy needs. Thus, effective, widespread public education on the benefits of renewable energy continues to be the cornerstone of any commercialization strategy. It's still necessary to create the climate—and the demand—for renewables. Since some of the greatest objections to renewable energy continue to stem from skepticism about its reliability, educating by example is a powerful mechanism. It's important for the public to see renewable energy in practice, to have their questions and concerns answered, and to be challenged to examine the underlying cultural beliefs and philosophy that make our country so energy-foolish.

## Conclusion

If you are reading this book, you probably believe as we do, that U.S. energy policy and practice need to be changed, and that greater renewable energy use is a promising path. You probably also believe in the power of change at the local level and may even be committed to initiating change yourself. If so, you will want to make sure that your time and energy are directed in the most strategic way possible for cultivating full-scale commercialization of renewable energy.

Much can be accomplished at the neighborhood and city level to encourage renewable energy commercialization. Local action is particularly effective if thousands of informed citizens are engaged in strategic projects all across the country. This kind of broad-based participation builds a groundswell of consumer demand that will be felt by utility companies, legislators, government administrative agencies, renewable energy entrepreneurs, those mysterious "market forces," and the public at large. Such popular demand, combined with policy action at the national and state level, will ensure that renewable energy use becomes standard practice in the United States. The stories you will read in the following chapters illuminate the origins of the groundswell of public support for a renewable energy future.

# CHAPTER TWO

# Strategy #1: Work in Partnership with Utilities

When only one person does something, sings Arlo Guthrie in "Alice's Restaurant," she's considered to be crazy. Three people do it and they're a conspiracy. Get enough people doing it, and you've got a movement that can't be stopped. Unfortunately, renewable energy is still in the conspiracy stage. There won't be a full-fledged movement until lots of us begin using renewables, and that means getting utility companies to seriously commit themselves to alternatives to fossil fuel.

Virtually every home and business in the United States gets its electricity from a public or private utility company. Only 250,000 Americans live or work "off the grid." So while individual actions such as installing photovoltaic panels on a summer camp help keep renewable energy technologies alive, it is up to consumers on-the-grid to convince utilities of the importance of those technologies, so that they will become healthy competitors in the energy marketplace.

Several utility companies have already made substantial commitments to renewable energy. Pacific Gas & Electric is aggressively testing photovoltaics for a variety of off-grid applications like warning lights on tall towers, illuminated highway signs, and gas and water meters. Idaho Power is offering a program to install and maintain photovoltaic power systems for homes and other applications distant from their distribution lines. Sacramento Municipal Utility District plans to install fourteen thousand solar water heaters atop customers' homes and businesses. Their work, along with that of a growing number of other utilities, has created a steady, growing market for renewable energy technologies.

But these are the exceptions rather than the rule. Most utilities continue to shy away from renewable energy, or make only a token effort for the sake of public relations. Some say they are "studying the situation," or "monitoring developments in renewable energy technologies," but have no action plan for the present or future. Others hesitate to explore renewable energy because of outdated information based on a number of companies' negative experiences with renewables in the 1970s and early 1980s. Why bother investing in new—and initially tricky—sources of energy such as wind, biomass, and the sun, so the thinking goes, when fossil fuels are easier and more familiar to use? Just as important, utility managers' planning methods or techniques are geared to consider conventional fossil fuel resources and don't readily handle renewable energy options. For example, whenever utility planners perceive a need to expand generating capacity—for instance, to handle projected population growth—they think about putting a large amount of capacity on line all at once, which is what would happen if a new coal-fired plant were added to the utility's generating mix. Renewables, however, are very flexible and can be added in small increments as additional capacity is needed. Most utility managers, unfortunately, are geared to deal with large, one-time, heavily capitalized construction projects, not small, ongoing investments, and it is difficult to change this orientation.

In the late 1970s, interest in renewable energy virtually exploded, thanks in part to the oil crises that forced U.S. drivers to wait for hours to gas up their cars and trucks. The Carter administration nurtured renewable energy technologies with its commitment to research and development—funding in this area rose from $75 million in 1975 to $718 million in 1980. Tax credits and incentives also helped push renewable energy into the market, sometimes prematurely. But the boom quickly imploded. One example of this is the collapse of the wind power industry as described by energy analyst Michael Brower in his book *Cool Energy*:

> The rush to build wind turbines brought many poorly designed machines to market that were either inefficient or unable to withstand the mechanical stresses caused by variable and high-speed winds; many were placed in less-than-optimal locations. Coupled with these technical problems, the reputation of the industry was seriously damaged by naive or dishonest operators overselling their products or seeking to take advantage of generous tax credits and a gullible public. These problems left a legacy of doubt and skepticism that is only now beginning to fade.*

*Michael Brower, *Cool Energy: Renewable Solutions to Environmental Problems* (Cambridge, Massachusetts: MIT Press, 1992), 71.

Nevertheless, a handful of the fittest companies survived, mostly those that offered reliable equipment and superior service. One domestic wind-turbine manufacturer, US Windpower, managed to hang on and ultimately design a turbine that some experts believe is making wind-generated electricity cost-competitive with electricity generated by coal or natural gas. Through a partnership with the Electric Power Research Institute and Pacific Gas & Electric, US Windpower's new 33M-VS wind machine is the first on the market that incorporates a turbine that spins faster in high winds and slower in low winds, along with advanced electronics that help this machine capture more power from the wind than its predecessors.

As renewable energy vendors are improving their technologies and making them more reliable, the whole energy playing field may be slowly tilting in their favor. Several states—Massachusetts, Nevada, New York, and California—now require utilities to choose new resources based on the *total* societal cost of generating electricity. In addition to the usual equipment, fuel, transmission, and maintenance costs, utilities must also factor in the environmental and health impacts of burning coal, oil, and natural gas. These "externalities," as they are called, usually appear as a cost per ton of emitted nitrogen oxides, sulfur oxides, or carbon dioxide. For renewable energy sources like the sun, wind, and biomass, externalities will be minimal or nonexistent. If the cost of externalities increased the price of coal-fired electricity by even 1¢ per kilowatt-hour, it would immediately make competitive the cost of electricity generated by wind at an excellent site. This kind of total-cost accounting will ultimately be an important factor in diminishing, or even eliminating, the price gap between fossil fuels and renewable energy sources.

Given the strides made by several of the country's utilities, others may merely need a little encouragement to follow their lead. Still others will only respond to steady pressure and clear evidence that investing in renewable energy can improve their bottom line. In both cases, citizen action is crucial. Few people relish such a task, and would place working with the local utility on a par with going to the dentist for root-canal work. Many utility companies may seem like monolithic, unfriendly, bureaucratic organizations that resist change, especially when it comes from the outside. Actually, state and federal regulations usually provide ample opportunities for citizens to influence power companies. All investor-owned utilities and some public ones must answer to the public utility or public service commission in their state. Commissioners often solicit input from groups representing consumers, environmental organizations, advocates for low-income renters and homeowners, and other constituencies whenever a utility seeks a rate increase, plans to build a new power plant, or develops a long-range plan. For example, a nonprofit environmental organization, the Conservation Law Foundation (CLF), has worked hard to carve out a role for itself as an advisor to utilities in New England. As a result, when the New England Electric System, a public utility holding company serving approximately three million people, developed its latest long-range plan, CLF participated as a full partner. Partly because of this collaboration, the plan's first goal is to "develop approaches to provide electric service in a more sustainable manner," and the plan emphasizes the development of renewable energy projects.

When it comes to municipal utilities—those owned and operated by a city—citizens have even greater influence over the decision-making process. These utilities answer to a board of directors, and those board members are usually elected or appointed from the community; many municipal utilities also have an independent citizen advisory board. Both of these bodies offer direct citizen input into the company's operations and philosophy. As an arm of city government, municipal utilities must respond to residents' concerns, and "sunshine" laws demand that their meetings and other activities be open to the public. Some are even required to put major projects to a public referendum. In Austin, Texas, for example, residents voted to dump the city's 16 percent share in a nuclear power plant after an eight-year, citizen-led battle (see Chapter 5.)

There's a right way and a wrong way for a utility to explore the bright new world of renewable energy. Blindly jumping in for public-relations reasons, or without a carefully drafted strategy for the long haul, can prove expensive, inefficient, and disheartening. This kind of disorganized approach can be a recipe for failure that delays a utility's eventual full acceptance of solar, wind, and biomass energy. A good model for introducing renewables is demonstrated by Waverly Light and Power, which provides electricity in northeastern Iowa. This municipal utility is taking a well-planned, systematic approach—first evaluating its current and future demand and generation options, then looking at both fossil and renewable alternatives for conservation and generation.

*Waverly, Iowa is a small town with a big vision: the Midwest's first utility-scale wind project. Renewable energy isn't new to Waverly, though—two run-of-the-river hydro turbines currently power the town's streetlights.*
Photo credit: Freeman Photography.

**WAVERLY, IOWA**     Waverly, Iowa, looks like television's version of a typical midwestern town. The picturesque Cedar River runs through the center of town, and tall trees line Main Street, which is graced with a turn-of-the-century, brick bank building and the landmark Waverly Theater. Some nine thousand people live in this growing farm town, which also hosts an expanding base of hard and soft industry, as well as the 140-year-old Wartburg College. Waverly is also the home of an extraordinary municipal utility, and boasts the Midwest's first utility-scale wind project.

Waverly Light and Power (WL&P) is one of the country's 1,981 local, publicly owned utilities.* Established in 1904, it serves roughly four

*A Small Town's Commitment to Conservation and Wind Power*

---

*These local, publicly owned utilities served more than sixteen million customers, and sold more than 291 billion kilowatt-hours of electricity in 1992. By comparison, 263 private power companies in the U.S. served more than eighty six million customers (84.4 percent), generating over 2.1 trillion kilowatt-hours of electricity (87.9 percent). These numbers are a bit "soft," however, because several sources provide conflicting information on the number of publicly owned utilities. Thus the slight difference between this footnote and the chart on page 28. Source: Energy Information Administration, 1993.

# WAVERLY LIGHT AND POWER TIMELINE
# FOR INSTALLATION OF WIND POWER

| 1/91 | 2/91 | 3/91 | 4/91 | 5/91 | 6/91 | 7/91 | 8/91 | 9/91 | 10/91 | 11/91 | 12/91 |
|------|------|------|------|------|------|------|------|------|-------|-------|-------|

|      |      |      | Feasibility Study |      |      |      |      | APPA grant for wind study ($17,690) |      | Installed monitoring equipment |      |

| 1/92 | 2/92 | 3/92 | 4/92 | 5/92 | 6/92 | 7/92 | 8/92 | 9/92 | 10/92 | 11/92 | 12/92 |
|------|------|------|------|------|------|------|------|------|-------|-------|-------|

|      |      |      | Installed monitoring equipment at Site 2 (3 mph improvement) | Signed land lease with Skeets |      |      |      | Pursued land lease for Site 2 |      |      |      |

| 1/93 | 2/93 | 3/93 | 4/93 | 5/93 | 6/93 | 7/93 | 8/93 | 9/93 | 10/93 | 11/93 | 12/93 |
|------|------|------|------|------|------|------|------|------|-------|-------|-------|

|      |      | APPA grant for wind turbine ($25,000) RFP for turbine | Selected turbine |      |      | Turbine delivered (7/15) Turbine erected (7/21) |      | Turbine on-line |      | Testing complete |      |

*There's a right way and a wrong way for a utility to explore renewable energy options. Waverly Light and Power is a good model of a company that takes a well-planned, systematic approach. This timeline identifies the major milestones in WL&P's quest for reliable, cost-effective wind energy.*

Source: *Quarterly Progress Report: Demonstration of Wind Turbine Operation for Utility Electricity Production in Midwest Wind Regimes,* R. Lynette & Associates, 1994.

thousand customers in a 33-square-mile territory. Since 1986, the demand for WL&P's electricity has grown almost twice as fast as the national average. Electricity sales have been increasing 4.2 percent a year, and peak summer demand rises 3.4 percent per year. Much of this increase is due to the city's aggressive plan to attract new business to Waverly.

Waverly Light and Power buys slightly more than half the electricity its customers need from Midwest Power System, a large, investor-owned utility based in Des Moines. It generates the remainder in several ways. Three run-of-the-river hydro turbines extract energy from the Cedar River, accounting for 0.5 megawatt—under 2 percent of the community's energy needs, but enough to light all its streetlights at night. Nine diesel generators, mostly used for peaking power, can crank out another 23.2 megawatts. Finally, the company owns a 1.1 percent share of a 650-megawatt, coal-fired generating plant in Louisa City, about 150 miles southeast of Waverly.

A general manager and a staff of twenty-seven run the day-to-day operations. Ultimate control and long-range planning approval rests with the board of directors. All are Waverly residents. Board chair Ivan Ackerman is a local attorney; vice-chair Lee Hinrichs and his wife, Pam, own a clothing store in downtown Waverly; Lois Coonradt is a local realtor; Christopher Schmidt teaches computer science at Wartburg College; and Ron Matthias is the dean of finance at Wartburg. None of the board members are energy experts, but all are creatively shaping Waverly's energy foundation and ultimately its future.

In 1990, the board published "2000 Points of Light: The Road to Excellence," a long-range plan for WL&P. The report identified fifteen conservation and environmental goals for the 1990s. These included appliance rebates, variable electric rates depending on time of day, and new home energy-efficiency programs. To back up its commitment to this plan, the board hired a new general manager to launch the plan's programs. Glenn Cannon had already made a name for himself in South Carolina as a creative manager committed to conservation. With this man and this plan, Waverly began positioning itself for a decade of innovation and responsible growth.

Cannon's first task was to kick off several energy conservation programs; and since conservation programs do not work without strong community support, Cannon promoted long-time WL&P employee James Jebe to the new position of Energy Advisor. Jebe started with the company as an apprentice lineworker in 1978, and worked his way through many of the company's field positions. As a native son of Waverly, he makes a better salesperson for new or unusual conservation programs than does an "outsider" like Cannon. As Energy Advisor, Jebe oversees the town's conservation programs, and helps devise creative solutions for people's energy needs. "If you're willing to spend the money, we're willing to help you become efficient," says Jebe.

Ten conservation programs are now running at full speed at WL&P. These include a local version of the national Good Cents program for new and existing buildings, which offers homeowners lower utility rates in exchange for meeting energy-consumption performance standards. These standards also encourage builders, architects, and homeowners to use energy-efficient furnaces, refrigerators, insulation, and windows, stimulating and creating a local market for them. Waverly Light and Power has also pledged $100,000 for a five-year tree planting effort. "Trees Forever" is coordinated and promoted by volunteer Kathy Sunstedt, an energetic Waverly junior-high-school teacher who has been involved in local energy issues for several years. The idea behind this effort is that planting trees around buildings substantially

*A strong commitment to energy conservation is the backbone of WL&P's long-range plan. A native son of Waverly, Energy Advisor Jay Jebe oversees the ten conservation programs now up and running, including the Good Cents program which offers homeowners lower utility rates in exchange for meeting energy-consumption performance standards.*
Photo credit: Nancy Cole.

reduces the summer cooling load. This award-winning program has another, more tangible benefit—trees add to the town's beauty and character. So far, approximately one thousand trees have been planted around homes and commercial buildings, with thousands more yet to be planted.

Intuitively, conservation programs make perfect sense. Reducing demand shrinks electricity use, which in turn reduces the operating cost for fuel or purchased power. Generating less electricity also means emitting less carbon dioxide, nitrogen oxide, and sulfur oxides, and depositing less soot and ash. In the long term, reducing demand can delay or help avoid the massive capital outlay for a new power plant. But energy efficiency programs, sometimes called "negawatt" programs, cost money today while the savings don't come until tomorrow. These investments often raise utility rates, which no one likes. But for those who participate in the conservation programs, reduced electricity consumption generally outweighs the increased rates, and participants benefit with lower monthly bills. Customers who don't participate in the programs, however, may see their bills increase slightly.

No matter how well WL&P's customers conserve energy, the company must still line up new sources of electricity. In 1999, the company's long-term contract for buying electricity from Midwest Power System expires. A new contract will undoubtedly mean higher rates. WL&P has several options: it can increase purchase agreements with nearby utilities, but this means uncertain costs and no equity invested in equipment. It could become a partner in the U.S. Department of Energy's new "clean coal" project plant in Des Moines, 120 miles southwest of Waverly, but the technology is unproven, its environmental credentials are shaky at best, and the costs are unknown. Several large utilities are tentatively planning a 200- to 300-megawatt coal-fired generating plant near Dubuque, about 100 miles east, but their plans may never materialize. What's more, each of these options sends energy dollars out of Iowa to pay for coal purchased from other states.

Wind power offered Waverly a more local, more independent option. While Iowa doesn't generally spring to mind as a windy state, several studies show otherwise. A national survey of wind energy performed for the U.S. Department of Energy shows strong potential throughout the Midwest. A recent study from the Union of Concerned Scientists called "Powering the Midwest" estimates that wind turbines could generate more than two trillion kilowatt-hours of electricity a year in Iowa. Even though Waverly doesn't lie in one of the state's windiest areas, Cannon and the board knew that wind turbines are getting cheaper and better, and that a handful of local sites are pretty breezy. So they put up several wind-measuring stations and hired a nationally known wind expert (R. Lynette and Associates of Redmond, Washington) to do a preliminary feasibility study. These activities showed that, in the right sites, wind turbines could reliably and cost-effectively generate electricity for the town. A utility-scale wind machine is currently operating on the crest of a farmer's hill 2 miles north of town.

The U.S. power industry is very interested in Waverly's wind dream, and is both supporting and following its progress. Half of the $35,000 for the initial studies and another $25,000 for the turbine itself came from an American Public Power Association's Demonstration of Energy-Efficient Developments (DEED) grant. What Waverly learns about wind energy could shed some light on non-California wind energy. To date, virtually all of the country's wind experience comes from the Golden State. No one really knows if a wind energy project in Iowa or Maine will operate like one in California, or even if the same kind of equipment can be used all across the country. Waverly's efforts will help other power companies establish projects wherever the wind blows hard enough.

*Glenn Cannon, WL&P's general manager, and Dr. Lou Honary, associate professor of industrial technology at the University of Northern Iowa, compare notes on the performance of the utility's first wind generator.*
Photo credit: Jeff Martin/UNI.

Even more interested in the project are Waverly residents. "People are always coming up to me after church or in the grocery store, asking me about our plans for wind power, and telling me they think it's a good idea," says Cannon. That's one of the advantages of a small municipal utility—the people who run it get constant feedback from customers. In Waverly, enthusiasm for the wind project ensures that Cannon and the board will keep it on the front burner.

As WL&P board members explored two separate energy paths for the future—conserving electricity and generating it from renewable sources—they looked for some way to organize all their options. They eventually commissioned an integrated resource plan. This handy tool, called an IRP for short, formally analyzes a company's entire energy

## WLP 1993 DSM PROGRAMS

**Residential**

Good Cents New Home

Good Cents Improved Home

Good Cents Home Loan Program

**Nonresidential**

Commercial and Industrial Lighting

Commercial and Industrial HVAC

Commercial and Industrial Motor

Commercial Audit

**Other**

Energy Efficiency Rate Structure

Trees Forever

Appliance Rebate

*WL&P spent 2.32 percent of its 1993 gross revenues on Demand-Side Management (DSM) programs—and avoided an estimated 2 to 4 percent of peak demand. By the year 2000, the Integrated Resource Plan (IRP) estimates a nearly 10 percent decrease in peak demand attributable to DSM programs.*

Source: *Analysis of Successful Demand-Side Management at Publicly Owned Utilities,* Oak Ridge National Laboratory, 1994.

profile. It evaluates current generating capacity, demand trends, future generating needs, and conservation opportunities. It then uses this information to identify which options will be the most cost-effective, both in the near future and years down the road. Such an analysis doesn't come cheap, and many companies shy away from doing one unless forced to by state regulations.

But the WL&P board realized that an IRP would be very useful in the decisionmaking process. What's more, in order to test the company's commitment to conservation and renewables, the board hired an independent utility consultant for the analysis who wasn't a strong fan of conservation efforts. By the time the consultant (named Thomas Wind, ironically) finished the study in June of 1992, he fully backed Waverly's conservation efforts and provided a road map for the utility's future. The IRP clearly showed that investing in energy efficiency would cost no more, and most likely cost less, than building new generating capacity. That's not always clear in a place like Waverly, where current rates for generating electricity are very low. (Electricity produced by burning coal now costs WL&P only 3.5¢ per kilowatt-hour.) But by looking to the future, the IRP pointed out that rates would rise, and by the year 2000 the town would probably face a generating crunch.

The plan evaluated thirteen conservation programs one by one. It showed that several of them should be started right away, others within the next five years, and still others delayed for at least five years, when the cost of purchased electricity is expected to begin rising steadily. According to the study, adopting all thirteen conservation plans would reduce peak electricity demand by 10 percent, and that 6 percent overall energy savings could be achievable with current technology. Assuming continued technology improvements, even greater savings can be achieved by the year 2010. Using this information, WL&P began the first ten programs.

Finally, the IRP pointed out that Waverly needed to add 4 to 7 megawatts of baseload generating capacity after 1999, in addition to pursuing aggressive conservation efforts. The study encouraged the town to begin exploring wind power for some of this capacity. Part of this conclusion was based on Waverly's access to a nearby site for reliable wind generation. It also had to do with the fact that the study assumed there would be a need to add 10 percent to the cost of electricity generated by fossil fuels to account for externalities. This narrowed the price gap between traditional generating capacity and wind power, making the latter a more attractive option.

Almost two decades of wind generation in California has helped work the kinks out of turbine design, operation, and maintenance. Researchers have learned how to design turbine blades, shafts, and other components that can handle complex stresses created by the wind. They've discovered that higher towers improve performance substantially. Several companies are developing new wind machines that analysts think will bring the cost of wind-generated electricity down to 5¢ a kilowatt-hour or less in the best wind sites within two to five years. This will make wind power competitive with coal-fired electricity. Wind projects employing from a handful to several score of these new designs are currently on the drawing boards in New York, Vermont, Minnesota, Illinois, Canada, and elsewhere.

R. Lynette and Associates' second, full-scale study examined two sites in WL&P's service territory. One was a plot of town-owned land near an existing transmission line. Several months of collecting wind speeds at both 20 feet and 140 feet above the ground revealed that the wind speeds were comparatively low. A field owned by farmer Russell "Skeets" Walther that sits atop a small rise proved to be a much better site. Wind speeds here averaged 11.7 miles per hour, fast enough for respectable energy generation from several different kinds of turbine. In addition, the daily variation in wind speeds correlated well with Waverly's demand for electricity—both peak in the late afternoon.

This analysis underscores the critical importance of considering local factors when evalu-

*Wind turbines at Altamont Pass, California.*
Photo credit: American Wind Energy Association.

## CALIFORNIA AND WIND POWER

The modern history of wind-generated electricity in the United States is really the history of wind in California. Thousands of turbines now spin along the ridges of the Altamont, Tehachapi, and San Gorgonio passes, churning out clean, reliable electricity. Over the past decade they have become part of the landscape, and now offer a dramatic backdrop for advertising agencies selling everything from automobiles to clothing. Today California accounts for 98 percent or more of the wind-generated electricity in the U.S., yet the state isn't even in the country's top ten when it comes to wind energy potential. According to "An Assessment of the Available Windy Land Area and Wind Energy Potential in the Contiguous United States," a comprehensive review of the United States' wind resources published in 1991 by the Pacific Northwest Laboratory, California ranks seventeenth behind such seemingly unlikely states as Nebraska, North Dakota, and even New York.

*Glenn Cannon and farmer Skeets Walther strike a deal. Eight short months later, in September 1993, WL&P's first wind turbine sits spinning on the small hill just behind Skeets's right shoulder.*

Photo credit: Nancy Cole.

ating renewable energy. Careful monitoring of the resource is essential. Windy corridors have their calm spots; low-wind regions have the occasional windy site.

Based on the second feasibility study, WL&P leased an acre of land from Skeets Walther. The 70-year-old farmer couldn't be a more enthusiastic supporter of the project. "I'm always interested in any project that helps us learn how to work better with Mother Nature," he says. He remembers, when he was a boy growing up on the farm, his parents used a small wind-powered generator to power electric lights, but that they gave it up when utility lines came to Waverly. He expects to continue growing hay and corn right around the first wind tower, as well as any others the utility may decide to build on his field.

In fact, the wind project is striking a responsive chord among many Waverly residents. Iowa is heavily dependent on energy imported from outside the state. That goes against the grain of Midwesterners, who are used to fending for themselves and who take pride in their independence. To the citizens of Waverly, wind energy symbolizes their do-it-yourself approach to life. Nancy Duneman wrote the editor of the *Waverly Democrat*: "For years I have hoped that our cities will try hard to meet their needs using wind or solar power. It makes me proud to live in Waverly. Many cities will look to us for guidance on this issue." Ike Ackerman says a woman from nearby Waterloo told him she wanted to retire to Waverly because of the town's search to reduce its dependence on fossil fuels.

Glenn Cannon encourages residents' enthusiasm but tempers it with a dose of fiscal reality. "People think that wind energy is free. There's no such thing as free energy. Our budget for 1993 included $125,000 for the wind turbine. For a small company like ours, that's a big commitment."

That commitment is to develop a conservative, well-designed plan for wind power. With $25,000 from the American Public Power Association, WL&P signed a contract with Zond Systems of California for an 80-kilowatt Vestas turbine that sits atop a 140-foot tower. This machine should provide enough electricity for about twenty average

Waverly customers. More to the point, it will give the company a chance to learn about this promising, home-grown energy option. And, by starting small, WL&P has the chance to change its mind without losing millions of dollars if wind power just isn't a good fit for the town.

In a way, Skeets Walther's field is an open-air laboratory for Cannon and his crew. They will learn how to operate, maintain, and troubleshoot a modern wind turbine. Working with a single wind machine means that the cost of electricity will be high. Without the DEED grant it would have cost about 9¢ per kilowatt-hour, which is two to three times higher than the cost of coal-fired electricity. But Cannon expects the cost will come down rapidly as turbine technology improves, as WL&P crews gain experience with wind power, and as the utility adds more turbines. He dreams of a small wind farm on Skeets Walther's land that could someday supply the utility's customers with more than 10 percent of their electricity. "My vision is 6 or 7 megawatts sometime in the future from wind," says Cannon. "That will fit nicely because we will have the staff and equipment to install and maintain that generation. The nicest part about it is that we won't be dependent on anyone or anything. Once it is installed it will be ours."

Wind offers Waverly many energy-generation advantages. It is modular, meaning the utililty can add a relatively inexpensive turbine or two whenever it has the money or needs more generating capacity. A coal-fired plant, by comparison, requires a massive investment all at once. Further, it offers a clean alternative to electricity generated by burning fossil fuels. "We have a moral obligation to our customers and society as a whole to provide electrical service in the most responsible manner, and that's got to include the environment," says Cannon. "We have our heads buried in the sand if we ignore that cost or pass it on to someone else."

Wind also presents some problems that will require creative solutions. A utility manager can't "dispatch" wind, meaning he or she can't turn on a wind generator whenever more electricity is needed. Generation occurs at the whim of the winds, so it can't be relied upon for peak power during the day. If wind contributes only a small fraction of Waverly's capacity, then it can be used to replace coal-fired electricity bought from Midwest Power System or elsewhere. But if it ultimately represents 10 percent or more of the town's generating capacity, then WL&P will need to develop other power sources that can be brought on line quickly when the breeze stops blowing. One option is putting existing electric loads—like hot-water heaters—under the utility's con-

| U.S. ELECTRIC UTILITY INDUSTRY IN 1992 | Number of Utilities/ Percentage | | Total Watt-Hours Sold/ Percentage | | Revenue in Billions/ Percentage | |
|---|---|---|---|---|---|---|
| Investor-owned Utilities | 262 | 8% | 2,099.9 | 76% | $148.52 | 79% |
| Publicly owned utilities | 2,017 | 62% | 386.8 | 14% | $24.44 | 13% |
| Federal power agencies | 10 | 1% | 221.04 | 8% | $13.16 | 7% |
| Rural electric co-ops | 943 | 29% | 55.2 | 2% | $1.88 | 1% |
| Totals | 3,232 | 100% | 2,763 | 100% | $188.50 | 100% |

*State and municipal utilities (publicly owned utilities) and rural electric cooperatives account for 16 percent of the energy sold in this country. They are particularly accessible to citizen input, and thus are good opportunities for renewable energy activism.*

Source: Energy Information Administration, 1992.

trol, so it can turn them off for a short time when generation from wind drops and while other generators are being fired up. Storing electricity or collaborating with an entire network of municipal utilities each producing its own wind power—another Cannon dream—could also help smooth out daily electricity generation.

Waverly Light and Power isn't stopping with wind, but plans to tap a host of renewable energy sources in addition to water and wind power. The company is exploring a methane-recovery system for a local landfill. Several WL&P staffers have converted a 1987 pickup truck into an all-electric vehicle, just to see how one would operate and be received in Iowa. Cannon is also studying biomass conversion and fuel cells, but plans to leave the research and development to larger utilities.

When it comes to wind, though, Waverly Light and Power *is* one of the large utilities. Its vision and pioneering work will not only prepare it to use this renewable resource, but will pave the way for others to do the same. "I would love nothing more than in the year 2000 to see Iowa with 50 or 100 megawatts of wind generation and know that Waverly was the first one out there," says Cannon.

CONTACT: *Glenn Cannon, General Manager, Waverly Light and Power, 1002 Adams Parkway, PO Box 329, Waverly, Iowa 50677; 319-352-6251.*

**LESSONS LEARNED**   There's no question that utility companies will be crucial to the growth and development of renewable energy technologies. Not only do they provide electricity or heat for the vast majority of U.S. residents, but they also wield substantial R&D budgets. A few utilities are already promoting photovoltaics and solar thermal water heaters in an effort to reduce peak demand and put off having to build expensive new power plants. Others could follow this lead.

No matter how large or monolithic a utility appears to be, citizen or activist input makes a difference. This input can range from simply writing letters expressing your interest in renewable energy, to asking questions about the company's plan for it during public rate hearings, to applying public pressure for your utility to engage in integrated resource planning that gives equal attention to energy efficiency and renewable energy options. If your utility is one of the leaders in promoting alternatives, express your support—letters from satisfied customers make for great public relations and marketing tools. If your utility is one of the laggards, make your voice heard in support of renewables.

In even the most fossilized utility there are people committed to change and renewable energy. Seek them out. These are the folks who will be most receptive to a plan for building a small wind farm or funding a photovoltaic project. They will also know how best to maneuver such proposals through the system. A local energy or environmental group will often know who these potential allies are. No matter what type of renewable energy project your utility may explore, citizen involvement can help ensure that your utility takes a well-planned, systematic approach that will succeed.

At the outset, it might seem impossible to discern what kind of project a utility might find interesting. Some may consider only big, showy, large-investment kinds, while others may prefer very small ones. Even small projects, though, can help a utility put its foot in the water, preparing—perhaps with a push from citizens and government—to take the plunge.

**SACRAMENTO, CALIFORNIA**     Some utilities must be coaxed, cajoled, and pushed into backing renewable energy or conservation projects. That's not the case in Sacramento, California, home of one of the most progressive utilities in the United States. The Sacramento Municipal Utility District, or SMUD, has long been a leader in renewable energy. The company's photovoltaic electricity generating plant, built in 1986 next to the now-closed Rancho Seco Nuclear Power Plant, is the world's largest at 2 megawatts. SMUD recently teamed up with Southern California Edison, another strong proponent of renewable energy, to build Solar Two, the largest solar thermal electricity generating plant of its kind. It is also testing small photovoltaic systems for

*Progress and Enlightenment at a California Utility*

*The Rancho Seco nuclear plant was shut down by public referendum, and several activists were elected to Sacramento Municipal Utility District's board, where they then promoted solar and wind power. SMUD now operates this 2MW photovoltaic electricity generating plant in the foreground of these nuclear cooling towers.*

Photo credit: U.S. Department of Energy.

powering everything from emergency road-side telephones to homes and charging stations for electric cars.

This commitment to both cost-effective renewable energy use and research and development stems from strong citizen participation and activism. It was a public campaign and vote that closed Rancho Seco. Several people active in that drive were later elected to the utility board, and helped direct its focus toward solar and wind power. The board hired a very aggressive pro-renewable and pro-energy-efficiency administrator, David Freeman, who declared as he took the job that it was his goal to make SMUD "the nation's leading solar utility within three years."*

SMUD also aggressively pursues energy conservation. The utility is looking to achieve an 800 megawatt "Conservation Power Plant" and 400 megawatts of renewable and advanced energy projects by the year 2000, through three main routes:

- Installing solar hot water systems atop fourteen thousand homes over the next six years. Not only would this substantially reduce electricity consumption, but it would also jump-start the market for these systems and bring down their cost.
- Offering financial incentives such as rebates and low-interest loans that encourage homeowners to replace electric water heaters with so-

---

*After three years at SMUD, Freeman moved on to head the New York State Power Authority.

lar water heaters, or to incorporate passive solar design in new construction and renovation.

• Investing in research and development of solar cooling systems.

"For a utility to be successful today, you need to go all out," says Donald Osborn, who directs SMUD's solar projects. "You must have upper management, the entire staff, and an enthusiastic public all working together to make sustainable energy a reality."

While your utility may not be as enlightened as SMUD, it may not be as resistant to renewables as you think. To find out where a utility stands on renewable energy and energy efficiency, call or write the public information department. For an independent view, get in touch with a local energy or environmental group—they often have valuable insights into a company's policies. With that information in hand, citizens can push a utility to seriously consider energy efficiency and renewable resources. Well-organized public support, pointing out the economic development and job potential at utility hearings and other public forums, can encourage such a shift. To make such arguments persuasive, they must be backed up with credible data and realistic economic projections.

Assembling a portfolio of successful utility examples can help sway doubtful utility executives. Supporting state and national legislation designed to accelerate the development and adoption of renewables is another tool for pushing reluctant utilities into action. Letters expressing your opinion never hurt, especially when they are also published in the local newspaper. That applies to congratulatory letters as well, applauding a new project or a risky stance.

CONTACT: *Donald Osborn, Program Manager, Solar Program, Sacramento Municipal Utility District (SMUD), PO Box 15830, Sacramento, California 95852-1830; 916-732-6679.*

**MORRISTOWN, NEW JERSEY**     Although clean, efficient heat pumps have been around for years, they haven't yet made a big dent in the home energy market. Part of the problem is their relatively high initial cost, compared to traditional heating and cooling systems, even though the energy they save can make up this difference in four to five years.

Jersey Central Power and Light's ambitious rebate program completely changes the financial picture for homeowners and makes heat pumps a cost-effective option right from the start. The company gives builders and developers who install ground-coupled heat pumps a re-

*Heat Pump Rebates at Jersey Central Power and Light*

*The Premier2 geothermal comfort system transfers energy from the earth to make heat, using electricity only to power a pump, compressor, and fan. Pictured from left to right are the heat pump, circulation pump, and water tank.*
Photo credit: WaterFurnace International.

bate. The rebate amounts to $400 per 12,000 BTUs per hour of heating and cooling capacity for installing the necessary underground heat transfer loops, and an additional $180 per 12,000 BTUs for heat pumps with an Energy Efficiency Rating (EER) above 13.* During the program's first year, fifty homeowners took advantage of the rebates and installed ground-coupled heat pumps.

Jersey Central Power and Light has been working closely with area builders and contractors, the people who are most likely to convince home buyers or builders to install these still-unfamiliar devices. It's a winning situation for everyone. According to the Environmental Protection Agency, ground-coupled heat pumps are the most efficient way to heat or cool a building. The earth coupling works especially well to isolate the heat pumps from severe winter cold or summer heat, so that these systems are applicable in most climates. This efficiency translates into reduced air pollutants like nitrogen and sulfur oxides, particulates, and atmosphere-warming carbon dioxide. Furthermore, the pumps help the utility reduce its peak and baseload demands during both summer and winter, delaying or preventing altogether the need for a new power plant.

CONTACT: *Jersey Central Power and Light, 310 Madison Avenue, Morristown, New Jersey 07960; 800-531-HEAT(4328).*

## Utility Incentives: The ZILCH and Energy Score Programs

**FORT COLLINS, COLORADO**     Citizen advisory boards were searching for ways in which the Fort Collins, Colorado, municipal utility could help homeowners cut their heating bills and improve local air quality. While identifying responsible energy practices for the future, the boards recognized that residents had no way to reliably compare the energy efficiency of two or more homes, nor did they always have the capital to invest in energy-efficient home improvements. So ZILCH and Energy Score were born.

Energy Score rates homes on a scale of 0 to 100—the higher the number, the more energy efficient the home. This independent but city-

---

*For ground-coupled heat pumps, the EER range is 10–19. An EER of 13+ means that about 4 BTUs of energy can be transferred to or from the ground for every BTU equivalent of electricity required to run the system.

certified analysis costs between $100 and $175, and the city contributes $50 of the expense. The analysis includes the size and shape of the home; its orientation; amount of insulation; number, size and placement of windows, doors, and skylights; air leakage; heating system and water heater efficiencies; and any active or passive solar features. The final rating is a far more reliable measure of energy use than, say, reviewing a previous resident's utility bills. If a home scores low on the rating scale, the owner can get an analysis of several options that will cost-effectively improve energy use. Energy Score has computer programs that chart the potential energy reductions and subsequent savings from the installation of better windows, a more efficient heating system, more insulation, or any combination of improvements.

After Energy Score makes its assessments, ZILCH—Zero Interest Loans for Conservation Help—comes forward with financial incentives. This program helps finance improvements suggested by the Energy Score analysis. A loan can be used to upgrade a heating system or install electric demand controllers, more insulation, a solar hot water heater, or a catalytic converter on a wood stove. It can also be used to replace an old polluting woodstove with a newer, low-emissions model.

CONTACT: *Lori Clements-Grote, City of Fort Collins Light and Power, PO Box 580, Fort Collins, Colorado 80522-0580; 303-221-6700.*

**BANGOR, MAINE** (Program terminated February 1994) Thanks to a partnership between Bangor Hydroelectric Company and Fleet Bank, central Maine residents were able to get no- or low-interest loans to weatherize electrically heated homes or to replace electric water heaters with solar or heat-pump water heaters. The no-interest loans were available to customers whose income fell below 175 percent of federal poverty guidelines; the low-interest version was for all others.

Interested customers started with a home energy audit from a Bangor Hydro energy specialist. This included an estimate of improvement costs, as well as a certificate of eligibility. The certificate qualified a customer for a loan application through a Fleet Bank branch. Loans could range up to a maximum of $2,000 for a single-family home or $8,000 for a multi-family building, and covered equipment and labor costs. Energy-conservation improvements had to be made within ninety days of loan approval, and were inspected by Bangor Hydro. Customers repaid these loans directly to the bank over a two- to four-year period.

*Low-Interest Loan Program*

Unfortunately, the program attracted only 200 takers out of a pool of 89,000 customers. One reason for the low response is that Maine residents tend to be quite loyal to their local banks, and view the very large interstate Fleet Bank with some suspicion. Moreover, since homeowners could only get a backup water heater, which would save them a modest amount compared to the $2,000 cost, few people felt it was worthwhile. The minimal customer response meant the program was not cost-effective for the utility, so it was dropped. This is a perfect example of a case where local activism could have made the difference between success and failure.

CONTACT: *Darrel Carter, Bangor Hydroelectric, PO Box 400, Bangor, Maine 04408; 207-945-5621.*

## City-Wide Energy Plan

**ASHLAND, OREGON**   The City of Ashland (Oregon) Municipal Utility has been promoting solar energy applications and energy conservation for more than a decade. Its wide-ranging programs cover a lot of ground. Ashland placed a solar access ordinance in its land use code, provides free home energy audits, offers grants for home weatherization, tests firewood for its moisture content (dry, seasoned firewood burns with less soot and smoke, thus reducing air pollution), and installs low-flow shower heads and faucet aerators. The city also helps builders and buyers design energy-efficient residential and commercial buildings, and provides financial incentives to encourage this.

Ashland's Energy Conservation Division, part of the city's Department of Community Development, has a staff of four to five people to carry out its mission. They work with developers and home buyers, sponsor community education efforts, and monitor the impact of the various programs. The city spends approximately $500,000 each year promoting energy conservation and encouraging renewable energy use.

As with most other municipal utilities, revenues from electricity sales fund a substantial portion of the city budget—approximately 30 percent in Ashland. The city's aggressive conservation program, however, has been matched by steady growth in demand, so revenues have remained stable. "The best thing about a publicly owned utility is that it has to operate in the best interest of the public," says a city spokesperson.

CONTACT: *Dick Wanderscheid, Conservation Manager, Energy Conservation Division, City Hall, 20 East Main Street, Ashland, Oregon 97520; 503-488-5306.*

**ELDORADO, NEW MEXICO**    The solar community of Eldorado, New Mexico began as a 1970s dream. Now it's a 1990s reality, thanks to strong support from the local utility. Virtually all of the 850 homes in this Santa Fe suburb were designed and built with passive solar heating in mind. The elementary school, library, and community center all rely on passive solar heat, as will planned office buildings. This is a perfect match between renewable technology and location: the region averages 310 sunny days a year; winters are crisp and cold, while summers rarely get hot enough to require air-conditioning.

In Eldorado, all energy needs are met by electricity, though this is an area where electricity rates are high, about 9¢ per kilowatt-hour. The Public Service Company of New Mexico realized that such high prices could scare customers away from electric heat, so it promoted passive solar homes with backup electric heat. The company distributed solar design workbooks to builders and potential home buyers. The company also encourages builders to install two electricity meters on each home. One measures total electricity consumption, the other records just the electricity used for heating. According to local builder Mark Conkling, owner and president of Sierra Homes, a typical 1,500-square-foot, three-bedroom, passive solar home costs about $60 per month to heat. That's roughly half the cost for electrically heating a similar, non-solar home.

Representatives from The Public Service Company developed the solar building design guidelines while working with students at both the local Luna Technical Vocational Institute and the University of New Mexico School of Architecture. These guidelines won the company a National Award for Energy Innovation from the U.S. Department of Energy. The company has also worked with builders and the Farmers Home Administration to create affordable passive solar homes.

CONTACT: *Rafael Tapia, Residential Marketing Team, The Public Service Company of New Mexico, Alvarado Square, Albuquerque, New Mexico 87158; 505-848-4681.*

**MEDFORD, MASSACHUSETTS**    To date, virtually all small-scale renewable energy projects have been aimed at residences. The first commercial application of photovoltaic-assisted lighting now brightens a Bradlees retail store in the Meadow Glen Mall in Medford, Massachusetts, thanks to a partnership that includes Bradlees, the University of Massachusetts at Lowell, and New England Electric Company.

*Solar Community*

*Bradlees PV Project*

An array of photovoltaic panels atop the store generates up to 4.5 kilowatts of electricity. The direct current is used by a bank of fluorescent lights fitted with energy-conserving electronic ballasts. During sunny midday hours, the PV panels can supply almost all the electricity for lighting the Junior Boutique, or approximately 3 percent of the entire store's lighting needs. Although this represents only a small share of the store's lighting, and the payback period is long, the project is important because it demonstrates the potential of PV-assisted lighting to the utility and commercial lighting manufacturers.

U-Mass Lowell got involved in this effort after learning about the direct current lighting system; graduate students built scale models to convince New England Electric it would work. New England Electric became involved because of its long history with photovoltaics. In 1988, the company installed PV panels atop thirty homes, a Burger King, City Hall, a fire station, and a furniture factory in Gardner, Massachusetts. This effort, designed to test the effectiveness of grid-connected solar panels, has been a success and serves as a model for others around the country. After making presentations to several potential site partners, U-Mass Lowell elected to work with Bradlees, which lent one of its stores for the direct-current lighting experiment.

CONTACT: *Bill Berg, U-Mass Lowell, Center for Sustainable Energy, 1 University Avenue, Lowell, Massachusetts 01854; 508-934-3377.*

*Joe Wiehagen (left), design engineer from U-Mass Lowell, and John Bzura, principal New England Power Service Company engineer, examine PAL's power panel. The Bradlees project is the only utility-connected PV system that puts solar energy directly into the lights, thus minimizing energy loss that usually occurs when the PV's direct current is converted to alternating current.*

Photo credit: Bill Berg.

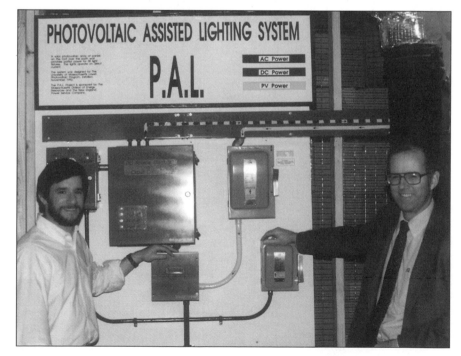

*These PV panels near Cooley Lake passively track the sun, powering the 10,000 Trees Project irrigation pumps in South Platte Park.*

Photo credit: Public Service Company of Colorado.

## LITTLETON, COLORADO

Trees lining a bicycle path built through the South Platte Park in Littleton, Colorado won't ever get thirsty, thanks to a photovoltaic irrigation system installed by the local utility, Public Service Company. Although it would only have cost $25,000 to extend electricity to pump water to the trees, the utility—with the help of the U.S. Department of Energy (DOE) and National Renewable Energy Lab (NREL)—opted to invest $41,000 for forty 63-watt PV panels that passively track the sun. The system pumps more than 10,000 gallons of water a day during the middle of the summer, delivering it directly to tree roots to minimize evaporation. Public Service Company chose the more expensive path in order to test the reliability of a modest-sized photovoltaic system, and so far it is working better than expected.

CONTACT: *Chris Thompson, 10,000 Trees Project, Public Service Company, 2701 West 7th Avenue, Denver, Colorado 80204; 303-571-3541.*

*Tree Irrigation Project*

## EUGENE, OREGON REGION

All across the country, long-buried and still-active landfills burp methane into the atmosphere. This energy-rich gas, generated by garbage-eating bacteria, is a potent greenhouse gas—fifty times better at trapping the earth's heat than carbon dioxide. The U.S. Environmental Protection Agency (EPA) estimates that landfills account for ten million metric tons of methane emissions each year.

That's not only an environmental danger, but a waste of a usable resource. A growing number of landfill owners and operators are

*Short Mountain Methane Power Plant*

*These engines burn landfill gas to create electricity at the Short Mountain Landfill gas-to-electricity project near Eugene, Oregon—a collaborative effort of the local power company and the county landfill.*

Photo credit: Emerald People's Utility District.

now joining with local utilities to capture this gas, clean it up, and burn it to power electrical generators. According to the Solid Waste Association of North America, collection systems currently trap and use methane from 114 landfills. Some sell the gas to a natural-gas utility or local business, others use it to generate electricity for sale to a utility or business. The EPA places the generating capacity of these landfills at between 315 and 350 megawatts.

The Short Mountain Methane Power Plant outside of Eugene, Oregon, is a good example of this kind of effort. A citizen's advisory

## NEAR-TERM BIOMASS OPPORTUNITIES: COFIRING AND REPOWERING

Many utilities are not currently in need of additional generating capacity to their system, but utilities can look for opportunities to use biomass in their existing infrastructure where it makes economic sense. In areas where large amounts of inexpensive biomass wastes and residues exist, utilities could modify the fuel-handling system at power plants so that biomass could be burned along with the existing fuel (cofired). Wood chips, for example, could be mixed with coal and then burned in an existing furnace.

Another opportunity for utilities would be to retrofit smaller, existing power plants to burn biomass in place of fossil fuels. Numerous small, fossil-powered facilities around the country were built more than 30 years ago, and many will soon begin to be retired. Utilities could look for opportunities to replace that lost power with biomass energy through renovating the retired facility to burn biomass (repowering). Overall costs may be much less for a retrofitted biomass plant than for a new fossil plant, and utilities might be able to gain attractive emissions credits for their system by utilizing biomass.

From *Biomass Power Commercialization: The Federal Role,* A Report of the Union of Concerned Scientists (May 1994).

board recommended that the local power company, Emerald People's Utility District, work with the county landfill operator to recover methane for generating electricity. The first phase of the project was completed in February 1992; an expansion, which doubled the project's capacity, was completed in November 1993. The $2.8 million price tag covered installing thirty-three gas collection wells, a gas cleaning system, and huge gas-fired generators. The project will generate twenty-six-million kilowatt-hours of electricity per year, enough to power seventeen hundred homes. As the landfill grows, the system will expand along with it, eventually peaking at forty-million kilowatt-hours per year. That's the equivalent of 87,000 barrels of oil, enough electricity for 2,600 homes.

Other cities or counties that collect methane from landfills include:

- Yolo County, California
- Raleigh, North Carolina
- Hartford, Connecticut
- Parlin, New Jersey
- Menomonee Falls, Wisconsin
- Kapaha (Oahu), Hawaii
- Littleton, Colorado

CONTACT: *Alan Zelenka, Senior Research Specialist, Emerald People's Utility District, 33733 Seavey Loop Road, Eugene, Oregon 97405; 503-746-1583 fax 503-746-0211.*

**PRINCETON, MASSACHUSETTS**    When the residents of Princeton, Massachusetts, discovered that their municipal utility was investigating buying electricity from the proposed, and highly controversial, Seabrook nuclear power plant in New Hampshire, they put their collective foot down. Since the utility already generated a significant fraction of its electricity from hydropower, residents urged it to look at another local, renewable resource—the winds that whip over the top of nearby Mount Wachusett.

Voters backed their lofty intentions with money, and approved a ten-year, $550,000 bond for a wind farm. Eight turbines were installed in 1984, and since that time have generated more than two-million kilowatt-hours of electricity. The wind farm currently supplies approximately 3.5 percent of the town's electricity. According to a Light Department brochure, "We remain dedicated to our wind site and are already planning to investigate an upgrade to a new generation of machine in the early 2000s."

CONTACT: *Sharon Staz, Princeton Municipal Light Department, 4 Town Hall Drive, Princeton, Massachusetts 01541; 508-464-2815.*

*Wind Farm*

**ROGERSON, IDAHO**     Chett and Kim Brackett finally silenced the dull diesel clatter that echoes over their 59,000-acre ranch 20 miles west of Rogerson, Idaho. For years they've depended on gas-guzzling generators for electricity to run the lights, kitchen appliances, and television

## PROFILES IN ACTIVISM
### Bob King

During an after-dinner walk on Thanksgiving Day 1972, 11-year-old Bob King stumbled upon the icon of his future—a long-abandoned hydropower plant on the Assabet River in Massachusetts. "I just knew then it would make sense to put old plants like that back to work," says King, now the president of SayWatt Hydro Associates. He recently refurbished an old hydropower plant on the French River in northeastern Connecticut. And he still owns the old dam on the Assabet that started his quest to develop small hydropower systems.

King followed a straight road from his boyhood interest to his current work. First stop was a mechanical engineering degree at Cornell University, which he chose because the school was renovating an old hydro plant behind its Syracuse, New York, campus. Next came two years with a San Francisco engineering company that specializes in hydropower projects. He worked on everything from a small 30-kilowatt hydro generator for an isolated village in Panama to a 6,000-kilowatt plant in New York State.

While driving through Connecticut's French River valley in 1986, King spied a sturdy but ruined powerhouse and dam. He knew right away he wanted it. The owner, a Boston company that had bought up a number of abandoned hydropower plants, offered King a job. Over the next two years, he negotiated with his new bosses, and eventually signed a $40,000 deal for the old powerhouse, dam, pond, and about

seven acres of land. In the spring of 1988, he set up a tent inside the musty, muddy powerhouse and started renovating the plant inch by inch.

King and his partner, Mit Wanzer, regularly invited dozens of friends to the river for weekend work parties. They gradually cleaned out the powerhouse, mucked tons of mud, slime, and catfish out of 150-year-old concrete canals, shored up the dam, and installed a new turbine. By the fall of 1990, the revitalized Nashaway Hydro Plant was churning out roughly 200 kilowatts of power, enough to supply the needs of about two hundred homes. King sells this electricity to Connecticut Light and Power—the company that abandoned the plant after the hurricane of 1938—for an average of 3.5¢ per kilowatt-hour. "The company then turns around and sells that electricity a few thousand feet away for 12¢ a kilowatt-hour," says a frustrated King.

The late 1980s and early 1990s haven't been the best time to develop small hydropower plants. An oversupply of generating capacity and the low cost of oil, coal, and natural gas have pushed the price of electricity way down. "The low avoided cost of electricity that big companies pay today is criminal," King points out. "Externalities would be the best thing that could happen to renewable energy, better than publicity or public awareness. Once we are all paying the real cost of energy production from fossil fuels, renewables big and small will be competitive."

Still, King can't help being enthusiastic about small hydro development. He offers five pointers for anyone interested in building or reno-

in a remote, turn-of-the-century house used by ranch hands. These generators also powered the pumps that bring up 5,000 gallons of water per day for the cattle that graze on their high desert land. The Brackett's ranch supports two thousand head of cattle, and nearly three hundred use the

*Bob King (center) and his partner Mit Wanzer (right) get a little help from some friends as they begin the back-breaking task of renovating the powerhouse at the Nashaway Hydro Plant.*
Photo credit: Betty Combs/SayWatt Hydro.

vating a small-scale hydropower plant and keeping the cost down:

- Pick the right site. Developing hydropower can subtly, or substantially, alter a river's flow. "You can't afford to anger local residents who think the project might endanger wetlands or ruin a nice stretch of river," King warns.
- Be prepared to do the work yourself. The Nashaway Hydro Plant was built with lots of sweat equity invested by King, Wanzer, and their friends. This saved the developers thousands of dollars in contractor and consultant's fees.
- Rely on sturdy, used equipment whenever possible. On the road near a working hydropower plant, King found two turbines which had recently been replaced with more modern ones. He bought them for $50 each. He built the Nashaway's generator from a twenty-year-old motor and got his gearbox from a paper mill. Total equipment cost: about $10,000. A new

turbine and generator would have cost $180,000.
- Figure out beforehand where you can sell the plant's electricity. Long-term contracts right now don't make much sense, since they could lock a plant operator into low prices for years. "If you can use the electricity yourself for a business or home, then you are basically running the meter backward and saving up to 12¢ a kilowatt-hour, rather than selling it for a few pennies a kilowatt-hour," King advises.
- Stock up on patience. Developing a small hydropower site requires working with dozens of federal, state, and local agencies. That means countless reports, inspections, permits, and an ever-growing pile of correspondence. Even building a simple 5-kilowatt generator on a stream running behind your house might take two years or more.

Is all the renovation work worth it? "I've loved doing it," says King with a big smile. But he's also itching for new challenges. One of these is an important reform project that might ultimately benefit people across the United States. King is challenging the conditions that stifle small power producers—who can never get fair rates as long as regulatory practices do not account for "externalities." If the social and economic costs of energy use are not accounted for, then nuclear and fossil fuels are valued artificially low, and small producers never get fair rates. Already making waves at the state level, King has testified in front of utility regulators and state legislators about more accurate ways to value and price energy. "You have to start small," says King, "and then who knows what can happen . . . ."

*The Bracketts' stock water pumping system is a good example of transitioning technologies. At the turn of the century, wind provided the power to run the pumps. Later, diesel engines were housed in the box at the turbine's base. Now, thanks to a new "energy service" provided by their utility company, the Bracketts use clean, quiet solar to water their livestock.*

Photo credit: Don Grundhauser/Idaho Power Company.

PV-pumped water at any given time. When the time came to extend power lines, which would have cost half a million dollars, the Bracketts chose to get their electricity from photovoltaic panels that are owned and maintained by the Idaho Power Company. Not only are things on the ranch much quieter, but Chett Brackett no longer needs to drive 30 miles a day to refill the water pumps' generators with gasoline.

Idaho Power Company's new off-the-grid electricity program illustrates the "energy services" concept that many utilities are adopting. It suggests that customers want light and heat, and power for refrigerators, televisions, and other appliances—energy services—and don't really care how it is delivered as long as it is cost-effective and reliable. For customers far from established utility lines, the company will design, install, own, and maintain a complete on-site photovoltaic system. The customer needn't fiddle with it, or worry if it breaks, any more than she or he would with power lines. The company charges the customer a set monthly fee that's independent of the amount of electricity used. That fee is 1.6 percent of the system's price.

Idaho Power Company currently sells residential electricity for about 5¢ per kilowatt-hour, one of the lowest rates in the nation. The photovoltaic systems cost six times more, in the neighborhood of 30¢ per kilowatt-hour. But when line-extension costs over Idaho's rugged terrain or ongoing fuel costs of gas generators are factored into the cost comparison, it is apparent that the photovoltaic systems make good sense financially.

CONTACT: *John Prescott, Idaho Power Company, PO Box 70, Boise, Idaho 83707; 208-383-2708.*

# CHAPTER THREE

# Strategy #2: Build Niche Markets

**S**cattered across the country, thousands of people live and work in buildings untethered to utility power lines. Some choose this option as a matter of principle, either because they wish to live independently or because they don't approve of the way power companies generate their electricity. Others simply live too far from the utility grid, or the cost of extending the closest power line is prohibitive. No one really knows how many people live and work off-the-grid, although the North American Electric Reliability Council estimates the number in the U.S. to be around 250,000.

Jim and Laura Flett are two such off-the-gridders. They own land in the foothills of northern California's Siskiyou Mountains. At first they lived in a trailer, hauling their water by hand and using kerosene lanterns for light. Then Jim rigged up a small electric generator from a lawn-mower engine and the alternator, voltage regulator, and battery from an old car. That gave the couple enough juice for a few electric lights. When the once astronomical cost of photovoltaic panels began dropping to earth, they bought two separate arrays: a couple of panels for the roof of the ranch-style home they had built, and two more to back up a not-always-reliable windmill that pumped their water. Today Jim, Laura, and their two children live much like their grid-connected neighbors in town. Twelve 60-watt solar panels provide more than enough electricity for fluorescent lights, a television and VCR, stereo, computers, kitchen appliances, and an efficient 12-volt refrigerator.

"I would probably take power from the electric company if they came out this far," says Jim. But he's not willing to pay the $25,000 that Pacific Power and Light would charge him for extending their lines. By comparison, his $10,000 investment in solar—and no monthly electric bills—make more sense. Jim sees several drawbacks to having the power company build lines further into the country: "I'd only want utility power

*Living in the remote foothills of California's Siskiyou Mountains, Jim and Laura Flett are "off-the-gridders." They get no electricity from the local utility; their household appliances are powered by twelve 60-watt solar panels mounted on the roof of their home.*
Photo credit: Laura Flett.

if they buried their lines. I wouldn't want to look at power poles for the rest of my life." He also worries that ready access to utility power could trigger a rash of home building in the rugged, isolated area.

The Fletts' use of photovoltaics represents what economists and energy analysts call a niche market—one of the many specialized and relatively small markets where renewable energy applications are currently cost-effective. The finances work out this way: In northern California, electricity coming out of a power line costs 5¢ per kilowatt-hour. Electricity from a new photovoltaic system costs much more, up to 30¢ per kilowatt-hour. On the surface, solar-generated electricity can't compete with electricity from traditional, central power stations. But add in the cost of extending the utility's distribution lines, and photovoltaics instantly become a cost-effective choice for this relatively small application—homes or businesses located more than half a mile from a utility grid.

Niche markets, so the thinking goes, help prove that new technologies like photovoltaics or advanced wind turbines work. They also spread the word about the technologies, and generate a steady and growing stream of orders that keep manufacturers in business. The concept didn't originate with renewable energy. To take one example, the ubiquitous air conditioner once cooled only businesses that required year-round cold. As the technology improved and prices fell, air-conditioning gradually became feasible for use in homes. Today, air conditioners are not only affordable, but have created vast new markets by making it possible for people to live and work more comfortably in the hot South and Southwest.

The same thing could easily happen with photovoltaics or solar hot water heaters. Someday city and suburban homes could sprout PV panels that would generate part or all of their daily electricity needs. That would mean cleaner air, more local jobs in a new domestic industry, and energy dollars staying within the community. Such a switch won't happen overnight, since photovoltaics at the moment simply cost too much. Exploiting a growing set of niche markets for photovoltaics helps produce what some utility executives and solar businesspeople call the "sustained orderly development" of solar electric technologies, forming a critical bridge between the high-cost PV systems of today and the widely affordable systems of tomorrow.

Thanks to the versatility of photovoltaics, remote homes represent just one of many potential niche markets. An Illinois company, for example, now makes virtually maintenance-free, PV-powered highway construction warning signs to replace noisy, expensive, diesel-powered models that need frequent refueling and maintenance. Solar panels even provide utilities with cost-effective electricity in some surprising tasks. San Francisco-based Pacific Gas & Electric (PG&E) now uses sunlight to power the warning beacons that sit atop transmission towers, power plants, and smokestacks, just a few feet above an enormous quantity of electric power. No lighting equipment can tap directly into this powerful flow—the lights would instantly burn out—so a transformer must be added in between to "step down" the voltage. (It's like trying to drink from a fire hose turned on full blast; you'd first need to somehow slow down the flow.) For a fraction of the cost of adding a transformer, PG&E can hook up a few PV panels, a controller, and a battery.

This last example is a classic niche market scenario for several reasons:

- The electricity running through the transmission lines probably costs the utility $1,000 per kilowatt to generate. Electricity from the PV panels, by comparison, costs closer to $5,000 per kilowatt. But throw in the transformer and the labor needed to install it, and this price gap flips in favor of photovoltaics.
- Since each solar-powered warning beacon is relatively inexpensive, installing several dozen carries little financial risk for a company like PG&E. So a small initial investment gives line workers, repair people, utility engineers, and executives an opportunity to test this new technology in a no-risk fashion.

- If the PV installation works, more systems can be added a few at a time. Once a number of systems are operating successfully and cost-effectively, then engineers will begin looking for similar sites where the technology can be applied.
- The market for photovoltaic panels to power transmission tower warning beacons is real, but limited—nothing that would shake PV panel makers into wild activity.

PG&E now uses photovoltaics in more than one thousand situations around its large territory. In many cases photovoltaics are simpler, cheaper, and safer than traditional generation, says Carl Weinberg, PG&E's former manager of research and development. Say you have a new house built a mile from the grid. Extending the distribution line to it costs about $25,000. "I could whip up a nice PV system complete with battery backup for that much," says Weinberg.

Other utilities have followed PG&E's lead. Solar panels are now the rule rather than the exception for several utilities. As photovoltaics are adopted by utilities across the country, PV manufacturers are seeing a steady increase in orders for solar arrays and batteries. In between the efforts of individual families like the Fletts and utilities like PG&E, are projects by communities and organizations to find niches where the application of renewable energy technologies makes sense. Two especially important projects are underway on the Hopi and Navajo reservations in the southwestern United States.

Many members of these Native American tribes live or work in buildings that aren't connected to a utility power grid. Yet they want to be able to have some of the conveniences of late twentieth-century life. They want electric lights so their children can read and study at night in the steady glow of these lights rather than the flicker of kerosene lamps. They want television, computers, and microwave ovens. As a result, some residents have turned to photovoltaics and soon, if ambitious programs blossom as expected, the two reservations could become the country's biggest proving ground for residential photovoltaics.

## The Hopis: Joining Tradition with Innovation Using the Power of the Sun

**ARIZONA**     From their new home, Janice and Joseph Day can see many of the Hopi Reservation's twelve villages, all clustered together atop three sacred mesas. The Days live on Second Mesa, next to the small arts and crafts store they own and operate along Arizona Route 264, one of the few highways that crosses the reservation. For years they have lived without electricity, even though utility power lines run right along the highway. "Solar looks better," says Janice, pointing to the utility poles. "People advise taking advantage of modern things that are good, but I don't think Arizona Public Service is good. You get dependent on it."

Now the couple enjoys electric lights, a small television and videocassette recorder, a radio, microwave oven, and a fan, thanks to three solar panels sitting atop a pole outside their home. They plan to add more panels to light their store, which they currently close when, as Joseph jokes, "it gets too dark to count the money."

Respect for traditional values and a distrust of outside agencies keep hundreds of Hopis off-the-grid. In fact, four of the twelve villages don't allow the local power company to extend its lines onto their land: Hotevilla, Walpi, Lower Moenkopi, and Old Oraibi, which is the oldest continually inhabited neighborhood in North America at nine-hundred-plus years. The leaders of these villages reject power lines for a number of reasons. They believe that energy emitted by power lines could disrupt the atmosphere and energy balance of the culturally important village plaza and ceremonial areas. They argue that power lines destroy the stunning panoramic vistas of the mesas, and must cross the land of people who will derive little or no benefit from them. Equally important, they worry that village residents will end up paying ever-increasing bills to an outside company. Autonomy is highly prized by the Hopi, who have never signed a treaty with the U.S. government.

On the other hand, even people living in the grid-banning villages would like to use modern appliances, watch television, and read by electric lights after sunset. The conflict between tradition and convenience intensifies as more and more Hopi children attend boarding school off the reservation. There they live in buildings with electricity wired into every wall and grow accustomed to conveniences that depend on this ready power supply. For many, returning to the village represents a loss of this freedom, and sometimes it influences them to move to the city rather than return home.

"Today it is important that people hold onto the old principles of Hopi while adapting to the realities of change," says one of the

The Hopi Solar Electric Enterprise has wired more than fifty homes on the Hopi reservation to take advantage of solar-generated electricity. Photovoltaics make a perfect bridge between the Hopis' desire to maintain traditional values and the need for modern-day conveniences. Strongly held cultural values also inspire the Hopi to help other indigenous people. Here, an Ecuadorian technician (right) gets a few pointers.
Photo credit: Owen Seumptewa.

elders of First Mesa. "By grafting the old and the new, present-day Hopi culture and society may continue to survive in a meaningful and purposeful way."

Photovoltaics make a perfect bridge, or graft, between old and new ways. The Hopi people have a deep reverence for the sun and honor it as the father of all living things, just as they honor the earth as the mother of all life. Catching sunlight atop a home and using its energy for light or cooking or relaxation fits this tradition perfectly. Former Tribal Chair Abbot Sekaquaptewa once compared photovoltaics to farming: "It's the same principle as when you raise corn and you gather the fruit of the earth. You're nursing from your mother." Many Hopis demonstrate this connection by offering prayer feathers to their photovoltaic panels, just as they do to their crops, rivers, the sun, and the earth.

Not only do PV panels directly use a traditional Hopi source of energy and sustenance, but they eliminate any need for an outside

*Solar panels on Joan Timeche's rooftop collect enough of the sun's energy to provide lighting in every room, power television and stereo equipment, and run the appliances in her kitchen.*

Photo credit: © The Arizona Republic. 7/22/91, Suzanne Starr. Used with permission. Permission does not imply endorsement.

power company with its intrusive poles and distribution lines. The Hopi's home-based electricity also turns out to be more reliable than utility lines, which are often knocked out by the heavy snows or high winds that sweep across the mesas. Sometimes during winter storms, "we like to sit at home watching TV and see the lights go out in the village," says Joseph Day. PV-powered electric lights are far safer than kerosene lamps, a common cause of fires on the reservation. And solar-generated electricity doesn't tie a family to monthly bills paid to strangers.

Through the efforts of the Hopi Foundation, a growing number of Hopi families are harvesting sunlight for their homes with small photovoltaic systems. This nonprofit organization, completely independent from both tribal and U.S. governments, was founded by four Hopis in 1984 to meet the needs of their community. Early projects included preserving several clan houses, the most ancient continuously occupied buildings in North America; developing a program for sustainably growing and marketing blue corn; and encouraging young Hopis to keep up the tribe's tradition of long-distance running. Then, in 1988, Foundation members started the Solar Electric Enterprise with the goal of bringing solar electricity to homes across the reservation. Seed money for the enterprise came from oil overcharge funds available through the state and from private organizations, including the Charles Stewart Mott, Hitachi, New Road Map, and Seventh Generation Foundations.

The first order of business was training several Hopis to install photovoltaic systems, wire homes for electricity, help residents maintain their equipment, and explain the advantages and limitations of this unfamiliar technology. Debby Tewa, Doran Dalton, Danny Humetewa, Doug Mapatis, and Harry Natumya spent several months at Appropriate Technology Associates (now called the Solar Technology Institute) in Carbondale, Colorado, learning PV wiring and troubleshooting. Now whenever someone on the reservation buys a photovoltaic system, the Solar Electric Enterprise team of Debby Tewa and Owen Seumptewa goes to work. Owen usually sets up the panels, ei-

*Hopi Solar Electric Enterprise PV technician Debby Tewa and her colleagues are "in training"—learning PV installation, electrical wiring, system maintenance, and public relations.*
Photo credit: Owen Seumptewa.

*Hopi Solar Electric Enterprise electrician Debby Tewa lives her work. Her home, shown here, was inherited from her grandmother and displays four PV panels on the roof and a bank of batteries around the back. Wiring and installation by Tewa, of course. Working with a colleague, installing a complete system takes about three hours.*

Photo credit: Owen Seumptewa.

ther atop the roof or on a pole next to the home, while Debby wires the house and rigs up the lights. "We meet at the battery box. The whole thing takes about three hours," Tewa explains. Homeowners regularly lend a hand hauling panels to the roof or setting the poles for a yard-installed system.

The Foundation's new headquarters, built on Third Mesa in 1991, is a standing advertisement for the solar power program. The building generates all of its electricity and hot water. A twelve-panel solar array mounted on the roof provides enough electricity for a bank of energy-efficient fluorescent lights, several computers, a fax machine, photocopier, microwave oven, coffee maker, and several rechargeable power tools. A new solar water heater provides 6 gallons of hot water an hour. People regularly stop by the headquarters to see photovoltaics in action. The Solar Electric Enterprise also spreads the message of photovoltaics through newspaper advertisements, a videotape, and flyers posted in chapter houses, offices, and stores. Every home that Tewa and Seumptewa wire up attracts curious neighbors and family members from both on and off the reservation. Enterprise members have also built a trailer complete with two 53-watt solar panels, four deep-cycle batteries, an inverter, and power controller. They will park it next to the home of anyone who's seriously considering buying a PV system and let them "test drive" solar power for a week.

This quiet campaign sells systems. So far, the Solar Electric Enterprise has wired more than fifty homes to take advantage of solar-generated electricity. The simplest one they advertise consists of a single 64-watt solar array with a battery, power controller, and inverter, which costs $975 installed. It can power two lights, a small television, and a videocassette recorder. Two-panel, two-battery systems cost $1,875.

In an area where more than 40 percent of the adults can't find work, and many others labor at subsistence farming, few families can come up with $1,000 to spend on a small photovoltaic package. The Foundation is therefore trying to reduce costs in every way possible. It buys solar panels in bulk to get the best possible deal, and passes the savings on to the new owners. Installation costs are kept low because Tewa and Seumptewa work for the Foundation, and only charge enough to meet expenses. The organization has also started a revolving loan program with a $50,000 grant from the Arizona Community Foundation. It loans residents up to $7,000 for four years at 8 percent interest. Such loans are expected to boost the demand for photovoltaic systems, since potential owners need not pay the entire cash price up front but can pay it off over four years.

In the coming years, the Solar Electric Enterprise hopes to become a self-sustaining business, rather than having to rely on grants and donations. "If we expect people to be self-sufficient, the project should be self-sufficient too," says Doran Dalton, chair of the Hopi Foundation board of trustees. This can be accomplished in several ways—selling more PV systems on the reservation, servicing systems no longer under warranty, or expanding the sales territory into the nearby Navajo reservation.

Dalton believes a compelling education campaign could convince those living in the traditional villages that solar power offers the independence that utility power denies, that it is cleaner and more respectful of the earth and its resources, and that it follows the Hopis' ancient traditions. Dalton estimates that the Solar Electric Enterprise could sell as many as two hundred photovoltaic systems to Hopi families who would like to have electricity but, in the spirit of their cultural tradition, want to remain off-the-grid.

**ARIZONA**     Just as their ancestors have done every spring for generations, Irving Benally and his family move from their winter camp in a sheltered valley to their summer camp on the high mountains of northwestern New Mexico. This Navajo family takes along its herd of sheep, its clothes, cookware, and tools. And for the last three years they've carefully packed along their photovoltaic panels and batteries as well.

Like the nearby Hopis, many Navajos have discovered photovoltaics as an alternative to kerosene lamps, smoky diesel generators, or utility power lines. Solar electric systems have been installed on more than thirteen hundred isolated homes over the past fifteen years, most of them bought and paid for with tribal or state funds. The Navajo

*The Navajos: Preserving Independence with Photovoltaics*

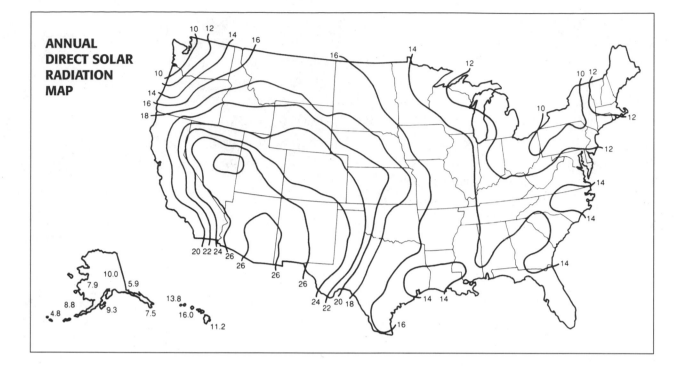

**ANNUAL DIRECT SOLAR RADIATION MAP**

*Using photovoltaic power on the Navajo reservation makes perfect sense: northern Arizona experiences brilliant sunshine year-round, and standalone PV systems can bring electricity to remote homesteads. This map shows annual direct solar radiation on a surface normal to the sun's rays, in megajoules per square meter per day. Note that the solar resource over one-third of the continental United States is as good as or better than that in southern California, where most solar development has occurred.*

Source: *Solar Resources,* Roland Hulstrom, 1989.

reservation, which is the largest Native American reservation in the United States, covers more than 24,000 square miles of rugged land in the "four corners" region of New Mexico, Arizona, Utah, and Colorado—an area almost the size of New England. Unlike the Hopis, who live in compact villages, the 180,000-plus Navajos tend to live far from each other, in dispersed family units loosely clustered by clan.

According to Larry Ahasteen, who oversees remote electrification programs for the Navajo Nation Government, more than ten thousand homes on the reservation don't have electricity. "Of these, the majority are in very isolated areas where electrical service is cost-prohibitive," he says. Stringing power lines to these remote homes would be a logistical nightmare and would cost millions of dollars, money neither the tribe nor residents have to spare. It's an ironic situation, since the reservation sits atop vast coal reserves and exports electricity from coal-fired generating plants to Las Vegas, Los Angeles, Phoenix, and other cities throughout the Southwest. Yet many Navajos living much closer to the source of this electricity can't tap into it.

They can, however, tap into a different, barely exploited renewable resource that's even more abundant than coal. The Navajos live in an area of brilliant year-round sunshine. Much of the reservation averages between six and seven "sun hours" a day, making it one of the best regions in the country for photovoltaic power. (Sun hours measure the average time each day that the incoming solar radiation measures 1,000 watts per square meter.) Detroit, by comparison, averages 4.1 sun hours daily, while Seattle manages only 3.7. In many parts of the reservation, then, sun-transforming photovoltaic panels can bring electricity to remote places where no power line would dare to venture.

A few of the early systems were purchased by individuals. Most, however, were bought by chapters, essentially small Navajo communities. Chapter leaders would ask the tribal government for financial assistance in getting photovoltaic systems for homeowners. The tribal government would then seek funding for a PV program, often from Arizona's share of oil overcharge funds.

Without some sort of grant, few families can afford even a basic two-panel system capable of operating a few lights and a small television. The initial cost barrier hasn't been the only problem. Many early systems were supplied by retail dealers who put up the solar panels, wired the house, and then left without fully explaining how everything worked or describing basic maintenance procedures. If something went wrong, these dealers or their representatives rarely made the long return trip for repairs. And few people on the reservation knew enough about photovoltaics to fix even relatively simple problems. Supplies or replacement parts were both scarce and expensive. To avoid such pitfalls, when the tribe purchases a batch of systems, the contract stipulates a two-year warranty and service period and also contains a training clause. So far, more than thirty Navajos have learned to install, maintain, and repair PV systems.

Since most PV systems operate for fifteen or twenty years, panels and batteries require regular maintenance long after the two-year service period. Many of the less technically inclined owners either forget or ignore this task. Rita and Cabby Yazzie live in a small timber-frame house 10 miles from the nearest paved road or power line. (The dirt road to their home is commonly referred to as a "washboard road," the kind that rattles a car even

*Navajo solar technician Dennis Jones provides routine maintenance to a PV system. Particularly cost-effective in isolated areas of the reservation, PV systems have another advantage—local employment. So far, more than thirty Navajos have learned to install, maintain, and repair PV systems.*

Photo credit: Larry Ahasteen/Navajo Housing Services Department, The Navajo Nation.

at slow speeds. From its juncture with the main highway, it passes no other homes or mailboxes.) Three pole-mounted PV panels feed electricity to three batteries, giving the Yazzies all the electricity they need. On a recent visit, Larry Ahasteen and Carmen Tsosie, who works for the Klagetoh chapter, found that the batteries were almost as dry as the land around the Yazzies' small home and traditional hogan. Ahasteen explained to the couple that they needed to keep the batteries filled with distilled water. Though Rita Yazzie nodded at the advice, she may not completely understand it or carry it out. According to Tsosie, this is a common occurrence all across the reservation.

Jimmy Daniels, manager for district operations at the Navajo Tribal Utility Authority (NTUA), may have a solution to this recurring problem. In essence he wants NTUA to put tiny utility power plants on homeowners' roofs, and sell them the electricity. The utility, which supplies coal-fired electricity to thirty-thousand customers on the reservation, plans to buy and install complete PV systems on forty homes that are at least half a mile from power lines. While they won't be the equivalent of a power line, these four-panel systems will give residents electricity for lights, a television, and other appliances. (They won't be able to power big appliances like refrigerators, though.) NTUA crews will maintain these systems and make all necessary repairs. The homeowners will pay a monthly fee, as they would if the power were coming from a utility line. With help from Sandia National Laboratory, the utility will monitor the performance of the systems and how well customers accept them. Other utilities are beginning similar programs across the country.

Ahasteen approves of this program because it could provide reliable electricity without homeowners having to maintain complex, unfamiliar equipment. But he's more intrigued by the Hopi Solar Electric Enterprise's approach, particularly because of its emphasis on self-sufficiency. "The Navajo people have become too dependent on handouts from the tribal and federal governments, and would benefit from greater independence," he says.

A separate Navajo effort could expand the role of photovoltaics in another niche market—water pumping. Every summer, the water table on the reservation drops so low that windmills can't bring water to the surface. Diesel or, if power lines are nearby, electric pumps must then take over the job. The Navajo Rural Development Corporation, a nonprofit organization that is not affiliated with either tribal, state, or federal governments, has developed and patented a solar-electric

pump that can tap into groundwater 1,000 feet below the surface. It is powered by eight 48-watt panels, half as many as competing solar pumps, and could cost about $4,000. Sunlight makes a great alternative to utility power, says Annaletta Osif, the organization's founder. "The power line is never yours. These pumps will belong to people. They will take care of them like their sheep," she says.

While getting electricity to Navajo families is an important problem, making sure they have comfortable housing is an even more pressing need. Ahasteen estimates that as many as twenty thousand of his people live in houses built to U.S. Housing and Urban Development standards that just aren't appropriate for the reservation. These poorly insulated cinder-block or timber-frame houses were not designed with the reservation's harsh climate in mind—100 degree F summer days followed by 40 degree F nights, and freezing winters with howling winds and heavy snowfall. Nor do they resemble the traditional Navajo home, called a hogan, which plays an important part in Navajo daily and cultural life.

A 1992 symposium brought together architects, designers, energy experts, Navajo community leaders, and tribal officials to plan comfortable, culturally appropriate housing.* Ahasteen calls the meet-

---

*The symposium's final report, called "Navahomes: A Dinelogical Study of Navajo Dwellings in Balance with the Land," contains a number of possible building techniques for Navajo homes. Many of the energy-saving techniques and designs could easily be applied elsewhere. The report is available from Larry Ahasteen, Project Manager for Navajo Housing Services Department, PO Box 2396, Window Rock, Arizona 86515; 602-729-4290.

ing "our housing pow-wow." The designs stressed a reverence for the landscape, traditional hogan architecture, use of indigenous materials, energy efficiency, and affordability. Framed homes insulated with straw bales and covered with adobe or stucco emerged as the best candidate. The first prototype has been built on the reservation near Ganado, Arizona. A greenhouse on the south-facing side provides passive solar heat during the winter, and a stove that burns pellets of wood and corn provides backup heat. Straw bales insulate the house three times more effectively than the insulation in the standard timber-frame homes built on the reservation. Electricity comes from eight roof-mounted photovoltaic panels. The cost for materials is estimated at $24,000, not including the PV system.

**THE COMMON THREAD**    Hopi and Navajo leaders use virtually the same terms when explaining why photovoltaics meet the needs of their people. "The Hopis are ecologists. We've always made do with what we had," says Owen Seumptewa, manager of the Solar Electric Enterprise. Larry Ahasteen echoes that sentiment: "The Navajos are master conservationists. Our traditional people teach us a lot about saving energy."

Both tribes share a reverence for the earth born from living in and adapting to harsh environments. Traditionally their members lived off the land, in rhythm with the seasons and in harmony with their surroundings. Until recently, Navajos and Hopis used only the energy they could themselves harness—the sun's warmth and light, the energy trapped in plants. But today, even many of those who follow the traditional ways also want to enjoy the benefits of life in the late twentieth century. And that usually requires electricity, something that until the last few years has been difficult for individuals to provide for themselves. Photovoltaic panels, however, offer even the most traditional individuals and families an acceptable bridge to modern ways.

The PV potential of both the Hopi and Navajo reservations could be a huge proving ground for this technology. Both from an individual and a utility point of view, such small projects could stimulate the PV market. Harold Post, project manager for the Sandia National Laboratory PV Design Assistance Center, explains that the two Native American projects account for approximately 1.2 megawatts, 6 percent of U.S. annual PV output. By itself this may not seem like a lot; but multiply this output by similar projects around the world, and it would make a significant contribution to PV commercialization. The

## SOLAR HOGAN PROJECT COLORADO

Government housing. The words conjure up images of plain, poorly insulated little houses scrunched together on long, treeless streets. From coast to coast, whether they are built in an inner city, a small town, or on an Indian reservation, they tend to look the same and fall apart quickly.

Charlie Cambridge has a different vision. On the University of Colorado campus he and architect Dennis Holloway have designed and built three different solar hogans, the traditional Navajo home. Doors face the dawn; cribbed, mud-covered roofs gently stretch up to the sky; and 90-degree corners have been banished (in Navajo lore, sharp angles can house evil spirits). These low-cost homes use energy efficiently, and need not be connected to utility grids, both important elements for homes designed for the sometimes isolated conditions of the Navajo reservation.

The smallest solar hogan is a one-room octagon, measuring 16 feet across. The walls are layered sandstone; the south-facing wall is glazed over to make a heat-absorbing Trombe wall. A central fireplace sits on the earthen floor, directly below a smokehole. Two other contemporary designs start with the traditional hogan and add more modern features. The largest is a three-bedroom, 1,600-square-foot home that also includes a kitchen and bathroom. Photovoltaic panels generate electricity, while solar collectors produce hot water and some supplemental space heat. Add a waterless toilet and propane-powered appliances, and the home can offer modern comforts far from the utility grid.

Cambridge and Holloway are convinced they can build modern hogans that Navajos

*Progress does not necessarily displace traditionalism, as Charlie Cambridge (left) and Dennis Holloway have demonstrated with solar hogans.*
Photo credit: University of Colorado at Boulder, Office of Public Relations.

can afford and will want to live in. To satisfy Department of Housing and Urban Development design and cost guidelines, however, they have had to replace locally available natural materials with standard lumber, plywood, and other store-bought materials. HUD approval is crucial for this project, since without it communities interested in building culturally appropriate homes won't qualify for Community Development Block Grants.

The solar hogan project has helped kindle interest in "culturally relevant housing." Several committed architects are working with tribes around the United States to develop home styles that combine energy efficiency with tribal culture and tradition. After Cambridge's solar hogan project was featured on the program "Beyond 2000," which was televised in more than one hundred countries, he received inquiries from all over the world.

CONTACT: *Dr. Charlie Cambridge, Navajo Hogan Project, PO Box 316, Boulder, Colorado; 303-494-9542.*

Hopi and Navajo projects are representative of those that could really get the PV market off the ground—small, stand-alone systems put in place around the world.

CONTACT: *Larry Ahasteen, Project Manager for Navajo Housing Services Department, Division of Community Development, Navajo Nation, PO Box 2396, Window Rock, Arizona 86515; 602-871-6999.*

CONTACT: *Owen Seumptewa, Manager, Hopi Solar Electric Enterprise, PO Box 705, Hotevilla, Arizona 86030; 602-729-4290.*

**LESSONS LEARNED**    The versatility of solar panels will help photovoltaics ultimately come into mainstream use. This will probably happen even though the electricity that PV systems yield, when calculated at the outlet, costs more per kilowatt-hour than electricity made by burning coal, oil, or natural gas. But, factor in all sorts of other costs, from extending transmission lines to the damages from air pollution, and in many cases photovoltaics already cost less in real terms than fossil-fuel–fired electricity.

There are many small niches in which PV can hold its own against traditional electricity. These include Navajos, ranchers, or vacationers, who own homes far from the grid; Hopis and others who don't want to depend on a utility for their electricity; and remote or small-scale applications like solar-powered streetlamps or highway warning signals. Interestingly, the day-to-day operations of utility companies represent tremendous potential volume for niche applications— from tower warning beacons to security lights for remote transformer installations. Each of these niche markets may produce only a trickle of orders, but many trickles ultimately make a rushing river. This flood will help manufacturers make panels at lower cost, which will further stimulate the demand. At some point in the future, the cost will be low enough that photovoltaic panels will make perfect economic sense atop homes already wired to the utility grid. But this future won't happen with PV—or any other renewable technology—unless currently cost-effective niches like those described below are fully exploited in the present.

## PV-Powered Highway Signs

**PENNSYLVANIA**    A Pennsylvania company has found a renewable way to eliminate a noisy, polluting roadside hazard—those diesel-guzzling warning arrows that steer motorists away from construction or obstructions. Protection Services, Inc., based in Harrisburg, Pennsylvania, makes solar-powered advance warners that could render their fossil-fuel–fired cousins obsolete.

Flashing-arrow advance warners powered by sunlight take advantage of a free energy source, emit no pollutants, require no maintenance, and operate silently, much to the relief of workers and nearby residents. Although they cost a bit more initially, they save approximately $1,900 in operating costs every six months. Diesel-powered models require regular refueling, oil changes, and maintenance; the solar versions don't. According to the company, their standard flashing-arrow board can operate continuously for as long as a month without getting any sunshine, thanks to energy-efficient, light-emitting diodes, which are used instead of standard bulbs, and to a quick-charging battery pack.

So far, Protection Services has sold more than six hundred of the PV-powered signs. The potential market is substantial—tens of thousands of these advance warners operate on U.S. roads every day.

CONTACT: *Craig S. Noll, Protection Services, Inc., 635 Lucknow Road, Harrisburg, Pennsylvania 17110; 717-236-9307.*

**GARRISON, NORTH DAKOTA** In the rangelands of the West and Midwest, many ranchers must pump water from deep wells to keep their cattle or sheep alive. Unfortunately, great pastures often sit far from power lines, making the cost of water pumping prohibitive. Photovoltaic panels are now popping up out on the range, offering a reliable, less-expensive alternative.

Charley and Clark Renfrew made this choice in 1991 after learning they would have to pay the McLean Electric Cooperative $4,700 to extend its lines to a pasture pump on their ranch outside of Garrison, North Dakota. For less than half that amount—$2,100—they installed a PV-driven system that pumps more than 700 gallons a day from their 80-foot well. The cooperative is testing a larger model capable of pumping 900 gallons a day from a 180-foot well that would cost approximately $3,200.

Being off-the-grid made PV-driven pumps a preferable option for Charley and Clark. This type of niche is still in its early stages for farmers on the grid and for farmers with large herds (the bigger the herd, the greater the water demand). Although PV pumps are currently custom-made, PV-driven farm equipment is rapidly moving into a new, cost-

*PV Water Pumping*

effective phase as researchers explore ways to create more marketable products.

CONTACT: *Dean Capouch, McLean Electric Cooperative, PO Box 399, Garrison, North Dakota 58540; 701-463-2291.*

## Solar Streetlights

**FLORIDA**   On first thought, solar-powered streetlights seem to make as much sense as screen doors on a submarine. At night there's no sun to power the lights. But there *are* batteries, which have been charged during the day by photovoltaic panels. In some cases, solar streetlights are cheaper than the traditional sentinels of urban nights. They also stand up better to hurricanes, as Florida residents discovered after Hurricane Andrew ripped through the state in the summer of 1992.

After howling winds demolished homes and downed power lines in Cutler Ridge, residents searched for their belongings by the light of solar-powered streetlights. In Kendall, people gathered at night under the lights to play cards and swap stories until electricity was restored to the community three weeks after the storm.

Solar Outdoor Lighting (SOL), a small company based in Stuart, Florida, has installed hundreds of its SolarPal lights in the state,

*With all conventional power lines down for as long as three weeks, solar-powered streetlights "saved the night" for several Florida communities devastated by Hurricane Andrew.*

Photo credit: Solar Outdoor Lighting, Inc.

and thousands around the world. They cost more than the standard, grid-connected model—$2,000 to $3,000, compared to $600—but cost far less to install. "If a community already has streetlights and wants more at shorter intervals, it is probably cheaper to use PV because it is expensive to dig trenches for power lines through existing asphalt and landscaping," explains SOL representative Alan Hurst. The Bent Tree community in South Miami installed twenty-eight SolarPals for approximately $80,000, saving an estimated $100,000 on installation and wiring costs for standard lights. Other niche applications include lighting in parks, parking lots, and along remote roads.

As the price of photovoltaic panels drops, these lights will become competitive even on urban and suburban streets. The panels require little maintenance, and don't rack up monthly electric bills. Nor does the electricity they use contribute to air pollution or global warming. A single SolarPal offsets the emission of more than 1,000 pounds of nitrogen oxides, sulfur oxides, and carbon dioxide each year compared to oil-generated electricity, according to SOL estimates.

CONTACT: *Alan Hurst, President, Solar Outdoor Lighting, 3131 SE Waaler Street, Stuart, Florida 34997; 407-286-9461.*

**ATLANTA, GEORGIA** In every American city, vacant lots provide overlooked opportunities for combining energy-efficient and renewable energy construction with affordable housing. The Cottage Home, designed by the Southface Housing Development Corporation, could fill this crucial niche. The Cottage Home is a small, single-family dwelling aimed at couples, single parents with one or two children, or single individuals. The houses cost between $40,000 and $59,000, depending on lot size and home design, placing them within reach of individuals or families who earn at least $12,000 per year.

The corporation is the building arm of the Southface Energy Institute, an Atlanta-based center for research and development of renewable energy and energy-efficient technology. Corporation developers interviewed potential home buyers to determine their priorities and needs, then used a team of architects, builders, and energy experts to design the prototype Cottage Home. The first home was completed in December 1991, in Atlanta's Edgewood neighborhood. Edgewood is an inner-city neighborhood with pockets of well-kept houses mixed with areas of rundown buildings. The basic one-bedroom Cottage Home measures 20 feet by 20 feet; the sleeping loft gives it 550 square feet of living space that feels like more, thanks to a cathedral ceiling on half the building and landscaping that makes it blend into the lot.

Not only do smaller homes require less energy for heating and cooling, they also use less material and energy during construction. In addition, Cottage Homes include a variety of features that further reduce energy consumption. These include airtight drywall, insulated doors and windows, passive solar design, and efficient appliances and lighting. The full range of energy measures adds approximately $2,600 to a home's cost, or about $300 a year to the mortgage. But they return almost $800 a year in energy savings, making the investment more than pay for itself.

CONTACT: *June Gartrell, Assistant Director, Southface Housing Development Corporation, PO Box 5506, Atlanta, Georgia 31107; 404-525-7657.*

*The Cottage Home: A Prototype of Affordable Housing*

## Solar Billboard for Auto Mall

**LANGHORNE, PENNSYLVANIA**     At night, the illuminated billboard advertising the Reedman Auto Mall in Langhorne, Pennsylvania, looks like thousands of others lining America's highways and byways. Daylight reveals that it's unique, powered by four photovoltaic arrays and a bank of batteries. "We never would have thought of using photovoltaics if the landowners had let us put in a power line," says Walt Bartkovich, maintenance manager for the Auto Mall. "But I'm glad we went with the solar panels. They've worked out to be much cheaper in the long run."

For $2,100, the company bought four PV panels and attached them to the top of the billboard. A bank of sixteen emergency light batteries stores energy from the panels and feeds the lights seven days a week from sundown until 2 A.M. Since the solar-generated electricity is free, the only additional cost is occasional maintenance and replacing the batteries every two years or so. The system has proved very reliable, and can light the sign even when skies are overcast for several days in a row. It has also proved very popular, so popular in fact that Bartkovitch has had to weld all the components in place so they won't disappear.

CONTACT: *Walt Bartkovich, Reedman Auto Mall, Route 1, Langhorne, Pennsylvania 19047; 215-757-4961.*

## School Crossing Lights

**PRINCE WILLIAM COUNTY, VIRGINIA**     Finishing touches were being put on the new high school in Prince William County, Virginia, when designers realized that no power lines extended to an intersection where state law required a flashing yellow signal. Local solar contractor Jody Solell recognized a potential niche for photovoltaics, called the school board, and was hired to build an 80-watt array to power the lights. It worked flawlessly. In fact, when Solell called the school board near the end of the year for some feedback, board members told him they would order another such system next time they built a new school.

CONTACT: *Jody Solell, Solar Electrics, 4060 Trapp Road, Fairfax, Virginia 22032; 703-425-9712.*

When niche markets are mentioned in the context of renewable energy use, photovoltaics applications most often spring to mind. However, any renewable energy technology that has relatively small-scale, localized applications can fit the concept. The main point is to create enough demand for specific technologies that the manufacturers can count on a steady increase in orders for their products. Steady orders, in turn, help decrease the product's cost and increase the product's reliability and visibility, thus leading to even greater demand.

The next three examples in this chapter explore different biomass applications—one fairly large, the other two small and local. All three illustrate specialized markets with great potential to advance the commercialization of biomass power. Biomass fuels currently provide approximately 5 percent of the nation's total energy demand, but much greater use would be possible if appropriate technologies and markets were developed.

Most biomass energy systems used in the United States today utilize low-cost wood residues, agricultural residues, or municipal wastes to generate electricity, heat, or steam. The fuel can be fed into a furnace to burn on a grate at the bottom of the firebox, ground to a small size and then blown into a furnace to be combusted, or burned as a layer of fuel supported on a cushion of air. Each approach has advantages and disadvantages for different types of biomass fuels and different end-uses. (For an in-depth discussion of biomass power, please see Appendix A.)

**PORTLAND, OREGON**    One of the most underutilized biomass resources in this country is sewage. Many cities and small communities collect sewage from homes and businesses, treat it, and then bury the resulting solid sludge in landfills. Several municipalities around the country are, however, looking upon this energy- and nutrient-rich material as a resource.

"The by-products of wastewater treatment—the gas, solids, and liquid—all have value," says Gene Appel, a principal engineer with Portland's Environmental Services Wastewater Treatment Branch. "They should all be used for the good of the community." The city's Columbia Boulevard Wastewater Treatment Plant is an example of how several by-products of sewage treatment can be used. During one step of this rather complex process, the bacteria that digest raw sewage give off methane gas as a waste product. At the Columbia Boulevard plant, this amounts to hundreds of thousands of cubic feet of gas per day, which is collected and burned. Some is burned at the plant to keep the huge digesters warm; some is sold to a nearby roofing company; the rest is flared away. Striving to be even more efficient in the near future, virtually all of the methane gas will be burned to meet other nearby heating needs or used to generate a portion of the fifteen-million

*Sewage Treatment*

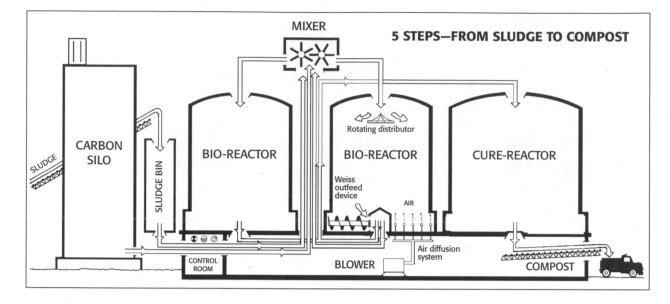

**5 STEPS—FROM SLUDGE TO COMPOST**

MIXER

CARBON SILO

SLUDGE

SLUDGE BIN

BIO-REACTOR

BIO-REACTOR

Rotating distributor

Weiss outfeed device

AIR

CURE-REACTOR

CONTROL ROOM

BLOWER

Air diffusion system

COMPOST

*Sludge is the residue that is removed from the city's waste-water as it goes through the treatment process. As the sludge decomposes, methane gas is created and then used as an energy source for the plant's operation. The composting facility is the last step in the sludge treatment process, producing nutrient-rich compost used in city parks.*

Credit: City of Portland, Bureau of Environmental Services.

kilowatt-hours of electricity that the plant uses each year. "Waste" heat generated by the electricity production could be salvaged (through a process known as co-generation) and used for the plant's space and water heating, as well as for warming the digesters.

In 1971, the City of Portland spent approximately $350,000 to construct a large sludge lagoon to hold the 30 dry tons of sludge it produced every day. Not only was the upkeep of this lagoon a substantial expense for the city, but it also took up valuable land space. So the city built a composting plant that turns sludge into a valuable fertilizer and soil additive. Some of this compost is used in city parks; the rest is sold to landscapers and individuals. Treatment plant managers are also exploring the possibility of pumping the nutrient-rich water left over after treatment onto nearby golf courses. It would both fertilize and water the grass, rather than being discharged into the Columbia River.

CONTACT: *Eugene Appel, Portland Environmental Services, 5001 N. Columbia Boulevard, Portland, Oregon 97203; 503-823-7185.*

## Energy from Manure

**MONET, MISSOURI**    On farms where cows or pigs are kept in feedlots, disposing of animal waste is a big problem. If the liquid waste is left on the farm, groundwater contamination can result from runoff; if the wastes are buried in a landfill, valuable nutrients such as nitrogen are lost. In addition, removing animal waste from the farm can cost thousands of dollars in time and transportation fuels each year. An experimental waste treatment system from the Barry (Missouri) County Extension Council helps reduce energy consumption while turning manure into a resource. Instead of hauling away cow manure, workers at the Nolan Meier Dairy Farm in Monet, Missouri, flush it into a specially constructed lagoon filled with fast-growing water hyacinth. The plants break down the manure and absorb its nutrients, growing rapidly. The resulting clean pond water is used to flush manure from the dairy barn floors. The plants can be harvested for animal feed or bedding, composted into fertilizer, or dried and burned as fuel.

According to calculations by Dr. Emmet Mc-Cord, a retired farm management specialist from the University of Missouri at Columbia, traditional manure disposal on a 120-cow farm costs approximately $8,000 a year. Besides providing an outstanding model of reusing materials, the flushing system at the Nolan Meier Dairy cut those costs by $2,700 a year. And using composted water hyacinths harvested from the lagoon for crop fertilizer offset the production of 24,000 pounds of nitrogen fertilizer, which would have required roughly 5,600 gallons of diesel fuel to manufacture.

This innovative waste management system would be useful on dairy farms by helping the farmer control waste and runoff. Research on waste management is being extended to testing this type of system for sewage operation in cities, small towns, and trailer parks where sewage leaches into the soil and contaminates the water supply.

CONTACT: *Dr. Emmet McCord, Route 1, Box 1860, Cassville, Missouri 65625; 417-847-2089.*

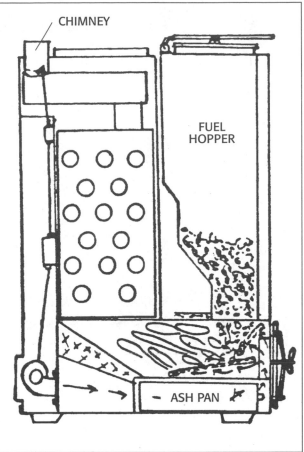

*A small Missouri firm has proven that home-sized biomass gasifiers like the one illustrated above are inexpensive to operate, using common waste items like sawdust.*

Credit: Raymond Rissler/Missouri Gasification Systems.

**CALIFORNIA, MISSOURI**   Biomass, once this country's most significant energy source, now represents less than 5 percent of the U.S. energy supply. Most of this comes from wood and paper companies that burn "waste" wood or paper for generating heat and electricity, though residential wood stoves account for a significant portion of biomass energy consumption. Current generation stoves reduce several problems associated with burning wood for heat. They are more efficient than old wood stoves or fireplaces, and generate less soot and air pollutants.

*Homemade, Home-Sized Gasification Systems*

Biomass gasifiers offer even greater efficiency and a cleaner burn, but are usually built for industrial-sized applications. In California, Missouri, a small town nestled in the Ozark Mountains, Raymond Rissler has been tinkering with home-sized gasification systems for almost a decade. Instead of directly burning fuel, his Missouri Gasification Systems heat wood, sawdust, or even old tires in the absence of oxygen, vaporizing a variety of flammable gases. These gases can then be burned in an adjoining chamber for heat or for generating electricity. They can also be trapped, condensed, and collected to be used later for running a truck or tractor.

## PROFILES IN ACTIVISM
### Debby Tewa

Debby Tewa lives with her feet in two worlds. In one she owns a traditional Hopi house without running water or electricity, attends ceremonial dances in her village, and speaks the Hopi language. In the other she wears blue jeans and T-shirts, works as an electrician, watches videotapes at home, and drives a bright red pickup truck. Through her work, Tewa makes it possible for more of her people to do the same: enjoy modern conveniences while staying true to their traditions and remaining connected to their community.

Tewa grew up in Hotevilla, a traditional Hopi village that still bans power lines. Raised by her grandmother in a sandstone and stucco house next to the village's ceremonial plaza, she learned tradition along with her ABCs. High school meant four years at a boarding school in California, followed by a trip to Flagstaff, Arizona, for college. But Northern Arizona University wasn't exactly what she'd expected, and she dropped out of her physical-education major, eventually returning to the reservation. While working in a job-training program for local teens, Tewa met a woman who was recruiting students for a Phoenix trade school. "Can girls

go too?" she asked, and was soon studying to be an electrician.

For nine years, Tewa lived in Phoenix, and worked at mostly commercial jobs wiring banks, businesses, and homes for small firms. While she missed her extended family in Hotevilla—"I'm an aunt to an 80-year-old woman, and a little boy is my 'boyfriend,'" she laughs—jobs on the reservation were few and far between.

One day in 1991, Hopi Foundation director Doran Dalton mentioned to Blanche Honani that he was looking for an electrician to work for the fledgling Solar Electric Enterprise, installing photovoltaic systems on the reservation. Honani just happened to be one of Tewa's aunts, and soon Tewa was back in Hotevilla, this time to stay.

Today, Tewa helps her neighbors turn sunlight into the electricity they need without becoming dependent on a "foreign" company, Arizona Power Service. She's become an expert at wiring small, traditional homes for electricity and connecting them to renewable, independently owned power plants—photovoltaic panels mounted on the roof or on poles in the yard. Her attention to detail and easy manner with people help ensure she's never at a loss for work. According to her Solar Electric Enterprise colleague, Owen Seumptewa, Tewa is highly respected by everyone,

Rissler uses his homemade gasifiers to heat a sawmill he owns as well as his home, and he has sold more than one hundred others in the United States, India, and the Philippines. His latest version includes an electric generator that yields about 15 kilowatts from about 15 gallons of sawdust, as well as waste heat that can be used for space or water heating. "The future is really good for something like this. It's not costly to operate, anybody can get the fuel, and there's not really any work involved to operate it," he says.

CONTACT: *Raymond Rissler, Missouri Gasification Systems, RD 3, Box 3612, California, Missouri 65018; 816-458-6511.*

*Debby Tewa, an electrician, helps her people enjoy modern conveniences while staying true to their traditions.*

Photo credit: Owen Seumptewa.

including people at Arizona Power Service. "They give her a lot of work," he says. "There are a lot of guys out here who are electricians, but Debby is better. She's more careful."

In addition to directly helping her people take advantage of renewable energy, Debby Tewa practices what she teaches. Her home, inherited from her grandmother, displays four photovoltaic panels on the roof and a bank of batteries around the back. Wiring and installation by Tewa, of course. She's also rigged up a solar shower, a rarity in a town without running water. It consists of a black, water-filled plastic bag set out in the sun. Absorbed sunlight heats the water, which Tewa pours into an insulated tank. A small pump powered by the batteries shoots water through a shower head strategically placed over a curtain-enclosed metal washtub. She's fixed up two others, one for her aunt and one for a co-worker, and hopes this could turn into a small side business.

In her spare time, Tewa volunteers at local schools and summer camps, teaching Hopi children and teenagers about solar energy. She's also travelled to Ecuador with members of the Center for International Indigenous Rights and Development. They worked in small villages teaching people to build solar ovens as an alternative to walking miles a day collecting scarce firewood. The next time Tewa goes to Ecuador, she hopes her Spanish will have improved so she can speak "Indian to Indian about problems facing native people in both countries."

The best traditions anchor us to our past while responding to our present and future, a continuous dance of old and new, as people like Debby Tewa help call the tune.

# CHAPTER FOUR

# *Strategy #3: Seek Creative Financing*

**S**unshine and wind don't cost a cent. Unfortunately, photovoltaic panels and wind turbines cost lots of them, an expense that prohibits many people from taking advantage of clean, local, renewable energy resources.

The high cost of renewable energy technologies has been described as a classic chicken-and-egg problem. Here's the chicken: low demand for wind turbines (or photovoltaic panels or solar water heaters) forces manufacturers to charge high prices for them. That is because the cost of equipment, salaries, advertising, and other overhead can only be spread out over a few, often laboriously made devices. Here's the egg: potential purchasers won't—or can't—buy these clean machines because they cost too much. Instead they wait, hoping prices will fall. But prices won't fall until demand rises, and demand won't rise until prices fall.

There are ways to break out of this cycle. Technology leaders, those brave, curious, or affluent souls who will buy an intriguing product even if the cost is high, can keep small markets alive. They also drum up further interest by showing off the new technology to friends, neighbors, and co-workers. If the product proves itself, and people decide they must have it, many will buy long before prices tumble. A few years ago, people happily parted with $800 for videocassette recorders that today sell for under $200. Others snapped up $3,000 computers that now cost less than $800.

On the other side of the checkout counter, companies may calculate that the potential market for a new product is so huge that they will hang in through several years of sluggish sales. The experience they gain and the reputation they develop will serve them well when demand begins increasing and prices begin falling. Many manufacturers of renewable energy technologies are in this position today.

Even though renewable energy technologies have been around for years, they are still stuck in this difficult young-market phase. Over the last twenty years, researchers and entrepreneurs have worked the kinks out of wind turbines, solar water heaters, photovoltaic panels, and biomass converters. No longer experimental projects, all of these are proven, reliable technologies. Trustworthy, committed manufacturers exist, and each year make limited numbers of their products. But while growing numbers of people are intrigued by renewable energy, high prices dampen their enthusiasm. Until costs come down, creative financing can help reduce capital outlays for prospective buyers, or at least spread them out over time. Strategies for financing the purchase of renewable technologies include outright grants from federal, state, city, and private agencies; matching grants; low-interest loans; investments from so-called green investors; buying serviceable used equipment; and donations.

Buying a $465,000 biomass gasifier, however, requires an entirely different approach. Leland and Gray Union High School in rural Townshend, Vermont, found itself spending $60,000 a year on electric heat. Switching to locally grown biomass (wood) would cut annual fuel bills by 90 percent, and save the district more than $1 million over a twenty-year period. Coming up with the necessary capital, though, presented a major hurdle, especially at a time when the local economy was faltering and people were putting off major purchases for their businesses and homes until things turned around. Yet thanks to a determined business manager, committed school-board members, and some very creative financing, the school managed to leap over the seemingly insurmountable capital barrier.

**TOWNSHEND, VERMONT**     Leland and Gray Union High School opened its doors in 1970 to students from the southern Vermont towns of Brookline, Jamaica, Newfane, Townshend, and Windham. The new building sported the heating technology of the day—electric resistance heaters. "At the time, the electric company was telling us that nuclear power was going to make electricity too cheap to meter," explains school board member Bill Koutrakos. But twenty years later the meter was spinning faster and faster, turning out heating bills that topped $60,000 in 1991. One estimate for the cost of electric heat in the year 2000 was $85,000; for 2010, $168,000.

In 1988, the school hired Frank Rucker as its business manager. The 28-year-old Montpelier, Vermont, native and his family had just

*Biomass to BTUS: How One District Went Back to School*

*Leland and Gray Union High School educates approximately 330 rural Vermont students, grades 7-12. The school saves $6,000 per year in heating costs because it switched from electric to biomass (wood chips) heat.*

Photo credit: Kevin Gallagher.

returned to the state after living in Dallas and New York for several years. After poring over the school's budget, Rucker realized that energy represented the single largest budget item behind salaries and benefits, and was the one he could most easily control. A quick comparison of local energy costs showed that electricity in southern Vermont cost $32 per million BTUs, oil about $7 per million, and wood about $3. "When I saw that as a business manager, I said, 'Holy mackerel! I'm operating at ten times the local baseline for fuel costs. What am I doing?'"

What Rucker *did* do was study his options. First he turned to the Institutional Conservation Program (ICP). This fifteen-year-old, federally funded but state-administered program helps schools, hospitals, local governments, and other institutions improve their energy efficiency. The ICP gave Leland and Gray $4,000 to audit its energy use and evaluate more efficient heating options. The school board matched this with another $4,000. In December of 1989, Rucker hired Vermont-based Enerwise Engineering to do the work.

Enerwise completed its report in March of 1991. It compared five different energy strategies for heating the school: coal, oil, natural gas, wood, and cogeneration. For Rucker and several school board members, the report presented them with just one sensible option—converting the school to wood heat. Their choice was firmly grounded in economics. A year's supply of wood chips cost one-tenth of a year's supply of electric heat and half the cost of oil. But other less bottom-line issues influenced their thinking as well. Buying wood from nearby suppliers supported the local economy, while buying oil or coal meant exporting dollars to another state. Wood also represented a manageable, renewable resource rather than an ever-more-depleted one. "When we put it all together, wood was the cheapest to operate and seemed to be the best bet. But it was the most expensive to install," says Rucker. The system he had his eye on, made by Vermont-based CHIPTEC, cost a whopping $465,000. A new oil-burning system, by comparison, would have cost $382,000.

Given annual electric heating bills of $60,000 and rising, the Enerwise report calculated that a wood chip gasifier would pay for itself in well under ten years, and ultimately save the town a tidy sum— $1,238,000 over a twenty-year lifetime when compared to staying with the existing electrical system, and $1,012,000 when measured against installing an oil heat system. After a relatively brief discussion, the school board asked voters in the district to approve a municipal bond of up to $420,000 for the conversion. Since several grant applications were pending at the time, the board told residents that the ultimate loan amount could be much lower. In April of 1991, official notices were posted, printed in the newspaper, and mailed to voters.

When the ballot boxes were opened, the overwhelming "no" vote shocked Rucker: "We thought that when we demonstrated the opportunity for savings, the dollars and cents would be obvious enough that people would show up at the polls, vote for the new system, and thank us for saving them $50,000 a year." Instead, many doubted the technology would work as claimed, and didn't believe the new system would really save them money. Irate and misleading letters to the editor that appeared in the local newspaper just before the vote also helped sway the outcome.

Fortunately for the taxpayers who support Leland and Gray Union High School, Rucker wasn't easily deterred. He kept Enerwise working on specifications for a new wood-burning heating system for the school, ostensibly because that was part of the initial contract. He also lobbied school board members, especially the three who made up the Building, Property and Energy Committee. The determined business manager invited Ralph Coleman, Bill Koutrakos, and Bill Castle— all long-time residents who were well-known and respected in the area—on a tour of the East Montpelier Elementary School, which had recently converted from electric to wood heat. After seeing this system in operation, and talking with the school's principal and custodian, they

*Rapidly rising energy costs were breaking business manager Frank Rucker's budget, so he cobbled together several different financing sources—a federal energy-efficiency grant, a state grant, a municipal loan, and the school board's reserve fund—to pay for the biomass system. Rucker's vision saves the school $6,000 a year, and will save more than $1 million over the life of the system.*

Photo credit: Kevin Gallagher.

became resourceful advocates for pushing ahead with conversion plans. "When we saw the Montpelier system, we really got turned on by it," says Koutrakos, a real-estate assessor from Windham who was initially skeptical of wood energy. "It's a nice clean system, with no piles of wood chips lying around."

Rucker, Coleman, Koutrakos, and Castle devised a two-pronged strategy for changing public opinion. On the financing side, the School Board asked the Institutional Conservation Program and the Vermont Department of Education to help pay for the wood chip gasifier. Both had the power to dole out substantial grants for energy improvements. Technically, however, the Leland and Gray application didn't meet the standard five-year payback rule, an important criterion for state funding.

In order to sell the wood-energy idea to voters, Rucker wanted to structure the financing so the school district would save money every year. That meant keeping the yearly debt service plus the $6,000 a year for wood chips below $60,000, the current cost of electric heat. And *that* required paying back the money over at least seven years. Once the new system had been paid off, though, the district's taxpayers would "earn" at least $54,000 a year—the $60,000 or more they would have spent on electricity minus the $6,000 they could expect to spend on wood.

Rucker petitioned the state's School Improvement Unit to waive the five-year payback rule. He used several compelling arguments:

- Leland and Gray was incurring "excessive energy costs," a condition that one state statute said would qualify a project for funding;
- A life-cycle cost analysis of the wood-heating system prepared by the Central Vermont Public Service Company showed that electric heat was the most expensive option possible, and that conversion to wood fuel was the least expensive;
- Finally, he cited a 1989 Executive Order by then-governor Madeleine Kunin requiring that agencies use rules "requiring that new construction or major renovation of such structures incorporate those practical energy-efficiency measures and energy-consuming systems that result in the lowest life-cycle cost . . . ."

Life-cycle cost analysis played an important part in Rucker's thinking. This approach measures the total life-cycle cost of operating a heating system (or any other kind of purchase, for that matter). It considers the cost of the equipment and interest on any loans plus the annual cost of fuel, labor, and maintenance, all in relationship to the system's life expectancy. The traditional method of measuring costs and savings merely considers how long it will take to save enough money to pay back the purchase price of the equipment. Of the two methods, life-cycle cost analysis presents a more realistic picture of the cost and savings of an energy system over time.

Doug Chiapetta, chief of Vermont's School Improvement Unit, agreed that this kind of analysis made perfect sense, and waived the five-year payback requirement. His ruling opened the door for Leland and Gray to receive a $140,000 grant from the state. The ICP added another $30,000, and the school board decided to use $30,000 from its reserve fund. That reduced the capital hurdle from $465,000 to "just" $265,000.

As Rucker negotiated for these grants, he and supportive school board members held four community meetings to discuss converting the high school from electric to wood heat. A few residents complained that this group shouldn't still be promoting the conversion idea after it had been soundly defeated at the polls; most, however, showed up to listen and ask questions. Using easy-to-follow transparencies and

*When Frank Rucker shows school district residents the CHIPTEC gasifier and controls they purchased, he also points to the school's monthly utility bills so they can actually see the savings.*
Photo credit: CHIPTEC Wood Energy Systems.

handouts, the wood proponents described the planned gasifier, and explained and re-explained the financing and how the system would actually save the district money. "It was a hard thing to sell because we were in some tight economic times," recalls Ralph Coleman. "Regardless of how much we preached that this was not going to cost them a cent, there were lots of people who just wouldn't believe that."

The presenters also fielded scores of other, non-financing, questions:

- Some residents were concerned that the wood gasifier technology hadn't really been proven. Rucker pointed to several Vermont schools and their successful conversion to wood, including the East Montpelier Elementary School. He also presented operating information from Minnesota-based Wood Energy, Inc., which has installed 150 wood gasifiers around the country, including one at the National Arbor Day complex in Nebraska City, Nebraska.

- Others asked about wood supply and price stability. Enerwise estimated that Leland and Gray Union High School would use about 200 tons of wood chips a year. That's less than three days of work for one of the five local wood chippers. According to Norm Hudson, the state's wood energy specialist, Vermont forests could easily supply enough wood chips even if every school in the state converted to a wood-burning system. What's more, only those low-value trees known as "culls" or "weeds" are used for chips. Removing them, says Hudson, actually improves the growth of remaining trees.

    Forests cover approximately 80 percent of Vermont's surface. The ready supply of wood, and thus wood chips, makes for extraordinary price stability. In 1980, Burlington Electric Company began buying wood for its 50-megawatt wood-fired electricity generating plant at a cost of $18 per ton. By 1992 the price had crept up to only $19 per ton.

- Pollution was another concern. Many Vermont residents use wood-burning stoves. They are all too familiar with the soot and smoke that wood burners emit, and the potential hazard from creosote fires in the chimney. Many also remembered the early 1970s, when the high school incinerated its trash and spewed ash and soot all over town. Wood stoves, fireplaces, and incinerators don't get hot enough to burn all the organic matter in wood or paper, explained Bob Bender, the CHIPTEC representative who attended some of the meetings. Unburnt material flies up the chimney as soot, or falls to the bottom as ash. The CHIPTEC system, on the other hand, heats wood chips to 2,300 degrees F, hot enough to burn creosote and other compounds that would resist oxidation in a standard 600 degree stove. According to company estimates, a standard wood stove or fireplace converts only about half of the wood it burns to energy. Up to 25 percent remains behind as soot and ash, and some 20 percent to 30 percent escapes from the chimney as creosote, unburnt hydrocarbons, or other pollutants. The gasifier, in comparison, is much more efficient, creates less than 1 percent ash and soot, and generates almost no hydrocarbons.

*To persuade school-board members to support the proposed wood-burning heating system, Frank Rucker set up a tour of East Montpelier Elementary School's successful operation. Principal Laura Johnson, above, described her school's dramatically reduced heating bills, while the custodian documented the system's easy operation and low maintenance.*

Photo credit: Kevin Gallagher.

- Residents also worried about increased truck traffic for wood deliveries, and pictured mountains of wood chips piled up behind the school. Not so, Rucker explained. He expected an average of one ten-minute delivery a month. The driver would back the trailer to a special enclosed shed behind the school and dump the chips into a hopper. From there they would be automatically fed into the gasifier.

For the real skeptics, the school board organized a field trip to the East Montpelier Elementary School to see a wood heat system in action. The visitors heard from Principal Laura Johnson, who told them her school's heating bills dropped from $25,000 a year to $3,600, even as the building's area grew by 50 percent. They also met with custodian George Matkowski, who praised the wood heat system's ease of operation and low maintenance.

After the final meeting in January of 1992, the school board scheduled a second vote. This time, however, the members wanted to avoid another anonymous ballot that offered no chance for them to rebut the last-minute, anti-conversion letters that were certain to appear in the local newspaper. So instead of a bond, they asked voters to approve a $265,000 municipal loan. The main reason for the switch was that loans are voted on during the school district's annual meeting. Much like the feisty town meetings held every March in each town in Vermont, the annual school district meeting is an exercise in direct, participatory democracy, providing a public forum for discussion and debate immediately before a vote. A loan also allowed Rucker to structure the repayment terms so the district would save several thousand dollars the very first year, compared to spending $60,000 or more for electric heat.

The loan strategy paid off. One resident opened the discussion by charging that "the school board is leading us into a barnyard that's filled with manure." Board member Ralph Coleman, known as a conservative maple-sugar producer, countered that charge. He explained once again the complex idea that raising and spending $265,000 would save the town $6,000 a year right from the start, and more than one million dollars in the long run. During forty-five minutes of debate and discussion, Rucker and board members fielded many of the same questions they had answered at the four informational hearings. They carefully described the financing, allayed fears over clear-cutting, soot, truck traffic, and messy piles of wood chips. When the loan was finally put to a vote, residents from the five towns approved it 168 to 85.

The opportunity to discuss the project right before the vote likely made the difference between victory and defeat, say Rucker and Coleman. Many of those who attended the district meeting had not attended any of the information sessions on the heating system. They expressed the same fears and concerns that had emerged before, which Rucker and the school board members were able to answer. Proponents of the new wood-heating system were also able to counter several inaccurate letters to the editor that again appeared in the newspaper the week before the meeting.

"Anything that relies on a public vote should be done in this [public] format," suggests Rucker. "Instead of showing up in a little cubicle and privately putting your vote in a ballot box, we had a debate and the opportunity to explain ourselves one final time."

Construction began in May of 1992. Installing the gasifier, the wood chip unloading system, and the automatic metering and delivery bins was actually the easy part. The *real* job was threading more than a mile of copper pipe throughout the school. The pipe carries gasifier-heated hot water from the boiler to classrooms, bathrooms, hallways, and offices. An electronic control system was installed at the same time. It links every part of the heating and ventilation system—the gasifier, boiler, each room in the school, and ventilation fans on the roof—to a central computer. The computer reads temperature information from monitors scattered around the school and automatically shunts hot water to wherever it is needed. By the time Leland and Gray Union High School opened in the fall of 1992, the new heating system was ready to go.

So far, it has performed better than expected, says Rucker. The gasifier required little maintenance throughout its first year of operation and far less attention from long-time custodian Wallace Holden than the previous electric system did.

The ash and soot problems that worried residents before the system was installed never materialized. Even on the coldest days, when the gasifier runs at full capacity, only a thin trail of clear vapor streams out of the school's chimney. And a year's worth of wood burning—211 tons of wood chips—produced just five barrels of ash, which residents collected to use as fertilizer on their gardens and maple groves.

The new system may actually save the school more money than expected on fuel because it uses energy more efficiently. In February of 1992, the old resistance heaters consumed 106,000 kilowatt-hours of electricity. In February of 1993, which was every bit as cold and windy, the new wood-powered heaters used the equivalent of 39,000 kilowatt-hours, a 63 percent reduction. Rucker attributes this savings to the gasifier's efficiency and the electronic control system, which has made the building far more comfortable with fewer hot spots and cold spots.

Ever since the new system was installed, area residents have been stopping by to take a look at the wood gasifier. All have been impressed, including some of the most vocal opponents. "I enjoy giving these tours to show people what they have bought," says Rucker. "I also show them our monthly utility bills so they can see how much we are saving already."

His perseverance has also generated some ripples around the state. The Vermont Department of Education changed its rules on payback periods and now allows grant applicants the option of using life-cycle cost analysis to prove a project's effectiveness. Rucker was appointed to the Vermont Superintendents Association's Energy Advisory Committee, which works with schools to make wise energy decisions. The committee is also drawing up a model school energy business plan to simplify this process. And Leland and Gray Union High School was recently used as the centerpiece of a short video promoting the use of wood in Vermont.

CONTACT: *Frank Rucker, Business Manager, Leland and Gray Union High School, PO Box 128, Townshend, Vermont 05353; 802-365-7355.*

**LESSONS LEARNED**     As Frank Rucker learned, there is money available for renewable energy projects. The trick is discovering the options, from traditional sources like the Department of Energy or the local utility, to less obvious ones like donations and green investors. The Leland and Gray story shows that sometimes you need to put as much, if not more, creativity into financing a project than into actually designing it.

Given the mind-boggling range of potential funding sources, don't quit when your project gets turned down. Present it to the next source on the long list you've developed by scouring grants books at the library, by asking environmental and alternative energy groups for tips, and by gathering suggestions from the organizations that express interest but decline to give you money. Perseverance is the key.

Successful funding also demands education. Sometimes those who hold the purse strings don't see the value of a particular project. Identify key decision makers, make sure they have all the information they need, and that they understand it. Frank Rucker knew that voters had the power to decide if his school should buy a biomass gasifier, so he laid out all the details for them .

Finally, new tools such as life-cycle cost analysis can help potential funders see a project in a new light. Spending $265,000 on a municipal loan for a heating system doesn't sound so bad when life-cycle analysis shows it will save more than $1 million over its lifetime.

The high capital barrier that blocks so many interesting and useful renewable energy projects won't shrink right away. Until it does, options like those described below can help.

## *Grants and Rebates Finance College Geothermal Heat Pump*

**POMONA, NEW JERSEY**     At Richard Stockton College of New Jersey, an aging heating, ventilation, and air-conditioning system, combined with budget cuts, helped the school take a calculated risk on a state-of-the-art geothermal heat pump. That system is now expected to save the state-run school an estimated $455,000 a year in operating costs.

College officials knew it was time to replace the twenty-year-old heating and cooling system. The question was, with what? One option was a gas-fired system, but with a price tag of $3.1 million, the financially strapped school could only have installed it in bits and pieces. Nor would it qualify for any rebates or grants. If the school found a system that would be extremely energy-efficient, the state EPA would kick in $2.3 million, and a state capital appropriation would add another $1.4 million. Plus, the local utility, Atlantic Electric, agreed to give the school a $1.2 million rebate if it picked a system that would help reduce the utility's peak demand.

These incentives helped Stockton officials buy a huge, $5 million geothermal heat pump using just $135,000 of the school's money. The system extracts energy from constant-temperature groundwater during the winter and dumps heat there during the summer. It is expected to cut the school's use of natural gas by 75 percent and reduce electricity consumption by 25 percent.

CONTACT: *Dr. Lynn Stiles, Richard Stockton College of New Jersey, Department of Administration and Finance, Jim Leeds Road, Pomona, New Jersey 08240; 609-652-4677 (for technological and research questions); Dr. Charles Tantillo, same address; 609-652-4381 (for financing and implementation questions).*

**CHADRON, NEBRASKA** Chadron State College found itself spending almost $300,000 a year on heat and hot water for its three thousand students. A 1989 study of several options showed that replacing the oil/natural gas heating system with a wood-chip–fired boiler would cut annual heating costs to $120,000. That's *if* the college could find $1 million for building the new wood-fired burner.

College officials made sure that state college board analysts evaluated the project's cost over its lifetime. This showed the wood burning system would save the college more than $1 million in heating costs over a ten- to twenty-year period. They also determined, through the U.S. Forest Service and Nebraska Forest Service that regional forests could sustain the college's demand for wood—about 2.5 million board feet a year. That meant the institution would be able to buy its energy source from a local supplier.

With this information in hand, the College Board applied to the state of Nebraska for funding from Petroleum Violation Escrow (oil overcharge) funds. In the 1980s, Nebraska had received money from penalties that oil companies were assessed for overcharging consumers during the price-control era between 1973 and 1981. The state then distributed some of that money for local energy-related projects. Chadron State College was awarded $1 million from these funds, and the wood-burning heating system was on-line by fall 1991.

CONTACT: *Suzie Shugert, Director of Public Relations, Nebraska State College System Office, PO Box 94605, Lincoln, Nebraska 68509; 402-471-2505. Physical Plant, Chadron State College, 1000 Main Street, Chadron, Nebraska 69337; 308-432-6202.*

**MISSISSIPPI** Every state in the Union has substantial biomass resources. Few, however, have catalogued them as carefully as Mississippi, or made it as easy for businesses and local governments to benefit from the survey. The diversity of the biomass resource is at once

*State Grant Funds Purchase of Wood-Fired Burner*

*Biomass Survey and Loan Program*

an important advantage and a drawback. The variety of fuel types makes it easy to match a wide range of applications. But too many choices can also baffle potential users, especially those more concerned with cost/benefit ratios than less-tangible advantages like environmental impacts or renewability.

The Mississippi Forestry Association and the Tennessee Valley Authority teamed up to measure the state's biomass energy resource. They listed current biomass users and fuel suppliers, surveyed the existing resource, and estimated its potential for development. All the data are included in the clearly presented report, *Biomass Potential of Mississippi*.

One of the survey's conclusions bears mention here:

> Locally, an industry or municipality may identify situations in which energy can be recovered from waste, while promoting economic development. *Even if the resource only has a relatively small potential nationally when discussed in quads, the contribution is major if it solves a waste disposal problem or results in more efficient usage of land, water, or economic resources* [italics added for emphasis].

While the survey alone meets an important need, the state went one step further. It established a loan program to help businesses take advantage of the biomass potential identified by the Forestry Association. The program offers loans of up to $200,000 at 2 percent below the prime rate for biomass projects that meet board approval.

The Jonston-Tom Bigbee Furniture Company in Columbia, Mississippi, is a successful illustration of the loan program's usefulness. The old wood-fired boilers at Jonston-Tom Bigbee were originally built in 1919 to power a steamboat on the Mississippi River. For the last forty-seven years they've worked on solid ground, heating and drying wood at the furniture company. When the boilers finally gave out in 1992, the company decided to replace them with new, more efficient, less-polluting boilers and a turbine generator. This combination would allow Jonston-Tom Bigbee to produce all the heat and up to half the electricity it needed each day. It also solved a major wood-scrap disposal problem.

Thanks to the biomass loan program sponsored by the Mississippi Department of Economic Development, the company was able to get a $200,000 loan for 2 percent below the prime rate. With another $400,000 of its own money, the company bought the turbine and generator. Today, the system produces between $10,000 and $15,000 worth of electricity each month, representing a four-year payback period. It could handle even more wood waste than Jonston-Tom Bigbee produces, and could theoretically make the company a net exporter of electricity. Unfortunately, electricity sold back to the utility would earn the company only a small fraction of what it has to pay the utility for electricity, according to CEO William Brown. If the utility ever changes those unfair rules, he adds, Jonston-Tom Bigbee could easily take waste wood from other nearby producers and generate a substantial amount of electricity for the utility.

CONTACT: *Wes Miller, Mississippi Department of Economic and Community Development, Division of Energy and Transportation, 510 George Street, Jackson, Mississippi 39202-3096; 601-359-6600. William Brown, Jonston-Tom Bigbee Furniture Company, PO Box 2128, Columbus, Mississippi 39704; 601-328-1685.*

**CALIFORNIA**    Drilling for geothermal energy can be a risky business. The earth's surface doesn't sport signs that say "Drill here," and, short of drilling, engineers haven't yet perfected a foolproof method for identifying sizable reservoirs of underground hot water or steam. But an innovative funding program in California now assumes the upfront risk of developing some of the state's renewable geothermal resources.

In a twelve-year effort, sponsored by the Research and Development Office of the California Energy Commission, the Geothermal Grant and Loan Program has distributed nearly $25 million for 150 geothermal projects—covering everything from research, planning, and development to the mitigation of any impacts associated with using geothermal resources. An award becomes an outright grant if a proposed geothermal project doesn't produce; successful projects, however, repay the funds to the Commission, thus rolling funds over to capitalize future geothermal projects. Commission funds are now available to both local governmental entities and private individuals or companies.

The citizens of Alturas (population 3,500) knew they lived in an area of northeast California with geothermal possibilities, and wanted to take advantage of this local resource. The Commission provided a grant to Modoc County to conduct a county-wide geothermal resource

*Geothermal Loans*

assessment. The assessment revealed substantial geothermal resources in the area, especially in Alturas. A subsequent feasibility study, carried out by the Geo-Heat Center—a research and consulting center at the Oregon Institute of Technology in Klamath Falls that was under contract to the Commission—determined that the high school in Alturas would make a good candidate for a geothermal project, since it was already heated with hot water and would thus require little retrofitting.

The high school well, drilled in 1989, delivers up to 80 gallons of 175 degree F fluids every minute to the system. It heats and provides hot water for the gymnasium/auditorium, several art and technology shops, maintenance shops, and a school bus barn. Encouraged by the high school's success, a second geothermal well is planned at the elementary and middle-school complex about a mile down the road. Once again, funding will be provided by the Commission and the school district. Tests indicate this well will provide more hot fluids than the schools need; the excess may be put to use as a mini-geothermal district heating system.

The two wells, pumps, heat exchangers, controls, piping distribution system, other equipment, and installation total $1.35 million. Commission awards cover 70 percent of the costs, while the school district supplies the rest. Currently, energy cost savings at the high school *alone* exceed $41,000 per year. This represents a simple payback period of about fourteen years; any price increases in conventional energy costs will shorten the payback period. One further advantage for Alturas: the funding agreement between the Commission and the school district calls for loan repayments for six years; after that, the district will realize the energy cost savings of the geothermal system for as long as it runs—which should be more than thirty years.

CONTACT: *Roger Peake, California Energy Commission, 1516 9th Street, Sacramento, California 95814; 916-654-4609.*

## Solar Equipment Rental Program

**SANTA CLARA, CALIFORNIA** Residents of Santa Clara, California, don't need any capital at all to invest in solar water heaters or pool heaters. Instead, they can rent the equipment from the city for a small monthly fee that includes system design, installation, and maintenance. Residents get lower water heating bills, and the city's municipal utility gets to reduce its electricity demand during peak hours. The goal of the program is to "keep local needs supplied by local services," says program manager Robin Saunders.

Between 1976 and 1981, more than two hundred residents took advantage of city-supplied solar water heaters. Since that time, however, few have taken advantage of the offer, since plummeting natural-gas prices make gas the less-expensive option. Once prices begin to rise again, the city is prepared to continue and even expand its solar incentive program.

CONTACT: *Robin G. Saunders, Solar Mechanical Engineer, City of Santa Clara, Water Department, 1500 Warburton Avenue, Santa Clara, California 95050; 408-984-3183.*

**VERMONT**    Energy-efficient mortgages provide a simple, affordable, and easy-to-administer tool for encouraging energy conservation. And in some cases they even help people with low incomes buy homes. Two basic types of energy-efficient mortgages exist. One blends money for improvements that will lower monthly energy bills into the home-financing loan. The other considers the reduced energy costs of an efficient home when calculating an applicant's income and expenses, making it easier to obtain a mortgage; less money spent on oil, gas, or electricity translates directly into more money that can go toward mortgage payments.

Generally accepted guidelines for energy-efficient mortgages have been in place since the early 1980s, and have been approved by the Veterans Administration, the Department of Housing and Urban Development, the Federal Home Loan Mortgage Corporation (Freddie Mac) and the Federal National Mortgage Association (Fannie Mae.) In 1992, the National Energy Policy Act authorized a pilot energy-efficient mortgage program sponsored by the Department of Energy. Any bank *could* offer an energy-efficient mortgage. Many don't, however, or at least don't advertise them. A well-organized lobbying campaign may be all it takes to encourage a local lender to begin offering these mortgages.

Many bankers say they would offer energy-efficient mortgages if only they had some way to clearly define an "energy-efficient home." What they would like is an independent, uniform, clearly understood system for rating homes similar to the way cars are tested, categorized, and compared by miles per gallon. A number of states, utilities, and nonprofit organizations are in the process of developing standard home energy rating systems that will eliminate the guesswork. There's even some progress on a nationally applicable certification process for individual state or nonprofit rating systems.

*Energy-Efficient Mortgages*

The nonprofit Energy Rated Homes of Vermont, for example, evaluates new and existing homes using an extensive checklist: home size, design, orientation, furnace and water heater models, insulation levels, placement and type of windows, lighting, results of a blower door test, and more. This information is run through a computer program that determines how many "stars" the home earns. Five stars indicate the most efficient home, one star an inefficient home. Four or more stars help a home buyer qualify for an energy-efficient mortgage, offered by several Vermont lenders. If improvements are needed to push a home into the four- or five-star category, their cost can be wrapped into the mortgage.

Energy Rated Homes of Vermont uses this example to explain the benefits: A one-and-a-half-story home that sold in 1991 for $88,000 needed $4,000 worth of work to change it from a two-star to a four-star home. The owners financed the work through their thirty-year mortgage at 10 percent interest, adding $37 to their monthly payments. But the improvements saved them $119 a month in energy bills, a net savings of $984 a year.

"I view the energy-efficient mortgage as an affordable housing tool that gets more people into housing and gives them more bang for their buck," says Janet Knutsen, a loan officer for Vermont's Chittenden Bank.

CONTACT: *Richard Faesy, Energy Rated Homes of Vermont, 7 Lawson Lane, Burlington, Vermont 05401; 802-865-3926.*

## *Partnership Creates Block of Solar Homes*

**PHILADELPHIA, PENNSYLVANIA**    A once-vacant block of land in an African-American neighborhood in North Philadelphia now boasts twenty-three affordable solar homes. A partnership between the National Temple Non-Profit Corporation (a community development corporation), the Department of Housing and Urban Development, and an initially reluctant local bank made this transformation possible.

Building affordable homes in urban neighborhoods is nothing new. Building *solar* affordable homes is. That's unfortunate, because passive solar design and daylighting can reduce heating and electricity bills, making it easier for residents to qualify for mortgages. "The lower heating bills would really make a difference for a hardworking guy with two or three kids who is barely surviving. It would mean more food in their mouths and clothes on their backs," says Larry VanStory, who bought one of the homes.

Each of the homes cost approximately $70,000 to build in 1985, and were sold for $30,000. Homeowners received a substantial federal solar energy tax credit (that is now defunct), and mortgages were subsidized through HUD's first-time home-buyers program.

The project won architect Robert Thomas and his company a 1991 award of excellence from The Foundation for Architecture in Philadelphia. "It's one of the most rewarding projects I've ever worked on," he says. The block faced south, perfect for passive solar design. And the community development corporation was intrigued by the idea that solar design could make homeowners more self-reliant. Both the architect and the Community Development Corporation had to spend time educating Providence Savings Association about the energy-efficient mortgage concept before the bank agreed to underwrite the home loans.

Vivian VanStory appreciates her home's high ceilings, great light, and low heating bills. She added a wood-burning stove, rarely needs to use the electric resistance heaters installed in each room, and spends no more than $90 a month on heating, even in the raw Philly winter. The southern exposure lets her have a great flower garden.

CONTACT: *Robert Thomas, Campbell Thomas & Co., 1504 South Street, Philadelphia, Pennsylvania 19146-1636; 215-545-1076.*

**GAY HEAD, MASSACHUSETTS**    When the Wampanoag Tribe of Gay Head (Massachusetts) decided to build a combined community center, meeting place, and tribal office on Martha's Vineyard, it turned to some very traditional sources for financing. The tribe asked the federal Department of Housing and Urban Development for help, and won a Community Development Block Grant that covered 20 percent of the $1 million building. The tribe itself put up $100,000. In a creative twist, though, leaders realized they might qualify for assistance from the Farmers Home Administration (FmHA). This federal agency usually funds projects in rural areas of towns smaller than twenty thousand people. But it also considers Native American initiatives, and the Martha's Vineyard site was rural enough to satisfy the FmHA. It approved a $750,000 loan at 5.25 percent interest over forty years. And though the agency initially balked at the tribe's plans for innovative energy systems and building materials, it ultimately backed them.

The new tribal center, designed by ARC Design Group in Chilmark, Massachusetts, honors the spectacular woodland site. It fits into

*Wampanoag Tribal Center Gets "Assist" from Federal Programs*

*Wampanoag Tribal Multi-Purpose Center was funded from the tribal treasury and two federal agencies—a Department of Housing and Urban Development (HUD) community development block grant and, in a creative twist, the Farmers Home Administration (FmHA). The Center's long southern exposure maximizes solar heating.*

Photo credit: Wampanoag Tribal photo.

the side of a hill, facing south for maximum solar heating. The builders, including Wampanoag tribal members, salvaged wood for interior framing from demolished buildings, wine casks, and elsewhere. They also used a variety of recycled materials, such as old tires for doormats, used windshields for tiles, and plastic soda bottles for carpets. Human waste is treated in a large composter and eventually used for fertilizer, while filtered graywater irrigates plants surrounding the center.

CONTACT: *Beverly Wright, Chairperson, Wampanoag Tribe of Gay Head, Box 137, Gay Head, Massachusetts 02535; 508-645-9265. John Abrams, ARC Design Group, PO Box 359, Chilmark, Massachusetts 02535; 508-645-2618.*

## *Factory-Built Solar Homes*

**PENNSYLVANIA**    An exciting alternative to site-built solar homes is now on the market, offering home buyers some welcome price breaks: factory-built solar homes. "These solar homes are affordable not only in terms of purchase price but also monthly bills," says Lyle Rawlings of FIRST, Inc., the research and design partner in a public-private partnership that aims to make energy-efficient solar homes in a variety of styles. The company has joined with the U.S. Department of Energy and Avis-America, a modular home manufacturer, to create the first factory-built solar homes.

"For the first time," explains an enthusiastic Rawlings, "a homeowner can simply go out and buy an integrated solar home. The solar and energy-conserving features are built right into the home in a con-

trolled factory setting, and shipped complete to the building site."
Styles under design development include one- and two-story ranches,
capes, and colonials. All will incorporate advanced passive solar design, high levels of insulation and tight construction, high-efficiency
lights and appliances, and photovoltaic panels for electricity. Buyers can
also select a variety of optional "extras" including an off-the-shelf solar water heater, or one can easily be added later to the solar-ready water piping system. Another popular customized feature is "earth tubes."
Plastic tubes are dropped deep into water or sewage trenches during
construction; a small electric fan then pulls air from the house or the
outside into these tubes. The system uses the constant temperature of
the ground to provide cool air in the summer and warmed fresh air in
the winter.

In eastern Pennsylvania, for example, a modular solar 2,000-
square-foot cape with a basement would cost approximately $150,000
including land, and the solar features would add between $10,000 and
$15,000 to the cost, increasing the monthly mortgage payment by
about $100. But utility savings should average at least $150 a month,
making the modular home cheaper to live in right from the start. The
first factory-built solar home, a 3,300-square-foot colonial, was recently
completed near Williamsport, Pennsylvania; six to ten more will be completed in 1994.

CONTACT: *Lyle Rawlings, FIRST, Inc., 66 Snydertown Road,
Hopewell, New Jersey 08525; 609-466-4495; Avis-America Homes,
Henry Street, Avis, Pennsylvania 17721; 1-800-284-7263.*

*The solar and energy-
conserving features of this
Bucks County, Pennsylvania
home are built right in at the
factory, significantly decreas-
ing the construction costs and
increasing the utility savings.
Participants in the Real Goods
National Tour of Energy Independent Homes get a good
look at the first of these
exciting homes.*
Photo credit: Lyle Rawlings, FIRST, Inc.

## "Green Investors" Back County Jail Solar Project

**ADAMS COUNTY, COLORADO** The Adams County jail uses 20,000 gallons of hot water every day to serve the needs of five hundred inmates. Industrial Solar Technology (IST) of Golden, Colorado manufactured a solar collector system capable of supplying industrial quantities of hot water, and the company's system now provides the hot water for the jail. Everyone is benefitting. "This doesn't cost [the county] a penny and saves them several thousand dollars" in energy and maintenance costs a year, says IST co-founder Ken May. It also keeps thousands of pounds of pollutants out of the state's air.

In 1986, for $150,000, IST built rows of parabolic troughs covering half an acre around the prison perimeter. Sunlight shining on each long, reflective mirror is focused on a black tube through which an antifreeze liquid is pumped. This hot fluid then heats water which is stored in a 5,000-gallon tank. A backup natural-gas heater switches on after dark, or when the solar system can't heat the water to 120 degrees F. The company sells this hot water to the jail for less than it would pay for natural-gas–heated water. From the roughly $7,000 it earns each year, the company keeps 8 percent to cover maintenance costs, and returns the rest to its investors.

The state energy office assisted in the project, providing funds to monitor system performance and to see how well the concepts of industrial solar water heating and private-public partnerships would work. The rest came from what May calls "green money," small investments of around $1,500 from friends, relatives, and others committed to financing environmentally sound projects. Most earned their investment back within five years.

*In tight times, state and local government initiatives to cut costs fit well with renewable energy technologies. Privately funded projects like this solar hot-water system at the Adams County jail—operating trouble-free for eight years—reduce both energy costs and maintenance expenditures.*

Photo credit: Warren Gretz/National Renewable Energy Laboratory (NREL).

Such a favorable payback period might not be possible today, May warns. In 1986, investors were able to take advantage of hefty solar tax credits, which have since disappeared. In addition, natural-gas prices have continued to fall, making it more difficult for solar thermal hot water to turn a profit. "We keep learning; the systems keep getting better, less expensive, and more reliable," says May. "With today's low energy prices, we're just hanging on and hoping the economic climate [for solar technologies] will improve." The company delivers solar hot water to other customers in California and Colorado, and is currently building a system to provide electricity to a remote village in Mexico.

Despite these difficulties, green investors are still out there, including those willing to wait for a ten-year payback. Furthermore, private projects like this one, which reduce energy costs and maintenance expenditures, mesh well with many city, county, and state initiatives to cut costs.

CONTACT: *Ken May or Randy Gee, Industrial Solar Technology, 4420 McIntyre Street, Golden, Colorado 80403; 303-279-8108.*

**WINCHESTER, KENTUCKY** A switch to renewable energy, especially when combined with an emphasis on energy efficiency, can help some businesses avoid the most difficult of financing decisions—bankruptcy or outright failure. When the renewable energy technology is purchased used or built with in-house labor and expertise, renewables are even more advantageous financially. Take the Freeman Corporation, a veneer manufacturer in Winchester, Kentucky. When Zeid Freeman took over the struggling business from his father, the company was spending nearly $300,000 a year on natural gas, and also paying a substantial sum to dispose of the 10 to 12 tons of wood waste it generated each day. This cash drain hampered the company's ability to compete and threatened its seventy-five manufacturing jobs.

Freeman realized the company's waste could also be its salvation. He bought a used pressure vessel for a wood boiler from a closed factory in Indianapolis, and studied European systems for burning "wet" fuel such as green wood chips. Company engineers then built their own chip- and sawdust-burning systems for $250,000, compared to over $600,000 for one from an established manufacturer. The scavenging effort was very worthwhile for the Freeman Corporation, because the biomass system provides all the steam heat needed to cook trimmed logs, dry veneer, or operate the veneer presses. Natural gas

*Used Products Save Money for the Freeman Corporation*

bills run less than $10,000 a year, mostly for space heating in the plant. What's more, the company is now producing three times more veneer than it did before the switch, and has won several international contracts. "Veneer companies that didn't [switch to biomass] aren't around

## PROFILES IN ACTIVISM
### *Rod Mercado*

Instead of chanting that once-radical mantra "power to the people," Rod Mercado actually helps individuals and entire communities acquire power—everything from solar power for cooking and heating to the power born from learning how to decipher an electric bill or fight City Hall.

Mercado has criss-crossed Texas with simple technology and simpler lessons. He's visited hundreds of ramshackle shanty towns (called *colonias*), big city slums, poor rural communities, and developments for elderly people living on fixed incomes. "We had a lot of misconceptions about what these people needed when we first started going in there," says Mercado. Then he laughs, explaining that "we" means "he and his white panel truck" with the University of Texas El Paso logo on the side.

Dollar for dollar, poor people benefit more than most from using renewable energy and making sure they use energy efficiently. Less money spent on heating and cooling means more money for food, rent, health care, or education. Unfortunately, much of the technology and information about renewables and efficiency have been developed for those with money to spare. Working with his colleagues at the University of Texas at El Paso, Mercado designed clear, easy, bilingual fliers and information sheets that take the mystery out of reading electric meters and offer energy-saving tips. He also collected information on "appropriate" energy technologies that people could fashion from common materials they might have on hand, like hand-made solar ovens and solar window boxes.

Mercado spread the word as he racked up more than 20,000 miles on his truck. He held informational meetings in schools, churches, and homes. By his own count, he has introduced thousands of people to the benefits of using the sun for free light and heat, and has taught them how to read an electric meter or water meter, in part to protect themselves against overcharges or fraud.

While on the road, Mercado depended on the kindness of strangers, and the "many, many good people who are out there." He also learned two valuable lessons about doing this kind of outreach work:

- Respect the people you visit, he urges, and ask them what they need. Then help them learn how to get it. This approach empowers people, in contrast to telling them what they need and giving it to them, which often makes them dependent.
- Don't go native, and assume you need to live in a colonia to help those who must live there. "It makes more sense to act as a bridge between the poverty that people are living in and the better life they are striving for."

Mercado's expertise in appropriate technology and alternative energy comes mostly from on-the-job learning. "You don't need to be an expert at something if you have the desire and the energy to learn about it and then get to work." He earned bachelor's and master's degrees in community planning and management, and

anymore," says Freeman. "It is one of the things which resulted in survival of this company."

CONTACT: *The Freeman Corporation, PO Box 96, Winchester, Kentucky 40392; 606-744-4311.*

worked on a variety of community issues for the city of El Paso for several years. One of the hats he wore was director of a seniors-helping-seniors weatherization program. That work put him in touch with the university's Energy Extension Service, now called the Center for Energy Resource Management. And the rest is road history.

As the energy program winds down, Mercado is setting his sights on an even more pressing problem—helping people who live in colonias and poor border towns get clean water for drinking, cooking, and washing. Contaminated water contributes to the high illness rate there, especially among children. The same straightforward approach that worked in the energy project may translate well to the water project: talking to men and women about how important it is to wash their hands before preparing food or feeding their babies, and teaching them how to make simple solar water distillers or composting toilets. More likely it will also take intense work at the state and federal levels.

The problem will get worse before real solutions are in sight, warns Mercado. Colonias are growing explosively on both sides of the border. Take Juarez, Mexico, for example. This city of 1.5-million people is separated from El Paso by the muddy width of the Rio Grande. Many of its inhabitants live in colonias of twenty thousand people. These unzoned, unplanned "neighborhoods" spring up with no running water, no electricity, no roads or schools. "For many people, seeing the colonias for the first time is like seeing the ocean for the first time," says Mercado. "Their size and their poverty just take your breath away."

*Rod Mercado's "rolling classroom" parks for awhile at the ARISE Center for Self-Sufficiency (Pharr, Texas). Women in the program display the solar box cooker they constructed.*
Photo credit: Rodrigo Mercado.

# CHAPTER FIVE

# Strategy #4: Change the Energy Rules

The early automobile didn't get much respect. In an 1899 editorial, the *New York Times* called it "unutterably ugly . . . a noisy and odorous machine." Other newspapers warned that horseless carriages threatened both public safety and public health. Some cities banned them inside the city limits, and New York State required drivers to pull over and shut off the engine if their idling auto made a passing horse skittish. Farmers' Anti-Automobile Leagues cropped up around the country to fight these noisy machines.

Anyone foolish enough, or extravagant enough, to buy one of these early automobiles faced serious logistical challenges. There were few places to refuel them and few mechanics able to repair them. Paved roads were a rarity—at the turn of the century, less than 200 miles of pavement existed outside of city limits, and 90 percent of country roads were little more than dirt tracks.

All that is ancient history. Today our lives, our cities, and even the geography of our country have been shaped by the automobile. Once little more than an amusing toy for the rich, automobiles are now perceived as a necessity for most Americans, rich and poor alike. They've been institutionalized in countless ways, from highway departments to the Strategic Petroleum Reserve to the industrial clout wielded by the Big Three automakers. The transition from "Get a horse!" to "Fill 'er up!" required vast changes in our social, economic, and political rules.

Renewable energy in the 1990s is like the automobile of 1900. People think of it as complicated, unreliable, expensive, and an extra hassle compared to the simplicity of using fossil fuels. Of course, there are numerous intriguing and successful renewable energy projects up and running around the country. But despite some projects' flawless operation for a decade or more, renewable energy is still regarded as a curiosity by most Americans. Just look at the numbers: in 1992, renewables other than hydropower contributed only 4 percent of the U.S. energy supply. Take out biomass and the contribution drops to 0.25 percent.

In order for renewable energy to play a much larger role, we need to demolish the barriers that currently limit its use. These include the public's misperception that renewables are a futuristic resource, the banking industry's reluctance to finance renewable energy users and entrepreneurs, a host of regulations that favor fossil fuels and hinder renewables, and a marketplace highly attuned to making a quick buck with no eye to the long term. Breaking down these barriers means changing social, economic, and political rules so they favor renewable energy, or at least put it on equal footing with fossil fuels and nuclear power. Such changes won't occur overnight, nor will they happen without a struggle. But they *can* be made by activists, community groups, city officials, transportation agencies, and utility managers. No sure-fire blueprint exists for forging such changes. One successful model is, however, beginning to emerge: first, create a favorable climate for renewable energy and energy conservation; then gradually move them from isolated curiosities into everyday practice; and, finally, institutionalize this shift. In a nutshell, that's what Austin, Texas, has done over the last twenty years.

In a state traditionally associated with oil, rugged individualism, and the gospel of bigger is better, Austin residents have turned their town into a national showplace for energy conservation and renewable energy use. What began as a small effort started by a few interested citizens evolved into the way the city does business. Energy conservation is now the major focus of a 150-person city department. The city-sponsored Green Builders Program won an award at the 1992 Environmental Summit in Rio de Janeiro, the only U.S. program to be honored there. The local utility operates two 300-kilowatt photovoltaic power plants and several smaller demonstration projects. Most important, the city's residents understand energy issues and are committed to innovative and efficient alternatives. "The average Austinite today believes that conservation is preferable to building a new power plant, that weatherizing a house is a good idea, and that you should be charged more if you use more electricity and less if you conserve it," says Peck Young, one of the architects of this change. "The value system that we helped put in place is a value system that won't be defeated."

Such a dramatic shift in values and priorities could happen almost anywhere. It requires solid community organization and a high level of citizen involvement. It also demands the courage to throw out old rules and make new ones. These can be as simple as changes in zoning rules or building codes, or as complex as creating Austin's Conservation Power Plant. Finally, success stories such as this underscore the crucial importance of combining grassroots activism with city politics.

## How the Tuesday Evening Lemonade Club Changed the City's Energy Rules

**AUSTIN, TEXAS**     The oil shocks of the early 1970s momentarily jolted Americans into thinking about energy. In Austin, they set the stage for major changes to the municipal utility, political structure, and citizen attitudes. At the same time that drivers waited in long lines for gasoline, natural-gas supplies were hitting all-time lows in the Southwest. That was bad news for Texas's capital city, which depended heavily on this fuel. The city-owned utility generated all of its electricity with natural-gas–fired power plants, while homeowners and businesses used it for heating and hot water. In December of 1973, supplies ran so low that the city had only one or two days of stored reserves.

That experience, along with forecasts of rapid population growth, forced Austin's energy managers and business leaders to think about diversifying the city's energy sources. They laid plans to build several coal-fired electricity generating plants. They also convinced voters, by a mere two hundred votes, to approve buying a one-sixth share of a planned 2,500-megawatt nuclear power plant, something they had rejected in 1971. This decision linked Austin with Houston, San Antonio, and Corpus Christi in building two reactors, South Texas 1 and 2, in Baytown, about two hundred miles southeast of Austin. Preliminary plans called for both reactors to be generating electricity by 1980, at a total cost of $900 million.

Despite these plans, Austin's energy future remained somewhat murky in 1975. The economy was booming as high-tech and biotechnology companies began springing up around the city. Demand for electricity was growing faster than expected, and officials worried that summer peak demand would soon outstrip the utility's peak generating capacity. The new coal-fired generators weren't scheduled to produce electricity until the mid-1980s, and trouble was brewing with the nuclear power plant. "The cost of that thing was rising just about every time you sneezed," says Peck Young. City officials, taxpayers, and anti-nuclear activists had good reason to worry: South Texas 1 and 2 ultimately cost $5.5 billion and were completed nine years late, in 1988 and 1989.

*The fruits of Austin's "energy revolution" are dramatically visible to residents and visitors alike: a 20 kW grid-connected PV system sits proudly on top of Austin's Convention Center. The City Council envisioned a "convention center of the future," promoting conservation and renewable energy technologies—the twin energy philosophies that Austin residents live by.*

Photo credit: City of Austin Electric Utility Department.

Against that backdrop, the story of Austin's energy revolution begins in 1976 with two modest, almost unremarkable events. Mayor Jeffrey Friedman set up a special panel to evaluate the city's electric rate structure. (Like a few other cities, Austin owns its electric utility. That gives city officials and voters the power to control the cost of electricity and, more importantly, the utility's investment and operating philosophy.) At the time, the largest consumers paid the lowest per-kilowatt rates while small users such as homeowners and small businesses paid the highest rates. Mayor Friedman's panel soon became the Electric Utility Commission, an official part of the city bureaucracy that would ultimately have a major impact on electricity use. Tampering with electric rates was a touchy issue in Austin, since the municipal utility accounts for about 44 percent of the city's revenues but only 29 percent of its expenses.

Also in 1976, the aide to a new city council member wrote a "white paper" describing the projected costs of building, operating, and ultimately decommissioning the South Texas nuclear power plant. Author Roger Duncan, a philosophy major just a few years out of the University of Texas, also detailed how renewable energy could provide a clean, inexpensive alternative to nuclear power. His paper eventually convinced the City Council to ask voters a second time, in 1979, if Austin should own shares in South Texas 1 and 2. Although they didn't change their minds, this vote got people talking about nuclear power and renewable energy.

**SETTING THE STAGE, SPREADING THE WORD**     Perhaps the most important step in Austin's energy about-face was a quiet, relatively common occurrence: people gathering to talk about a topic near to their hearts. At a popular health-food store, two groups began meeting separately each week to talk about energy issues. Roger Duncan and Peck Young had co-founded the first group, Austin Citizens for Economical Energy, with the intention of getting Austin to reverse its commitment to nuclear power. This was quite a turnaround for Young, who had managed the pro-nuclear power plant campaign in 1973. "I believed in nuclear power the first time around. But when I realized no one had any idea how much that thing was going to cost, I couldn't do anything else but fight it."

An alternative energy discussion group was also in full swing. Its members included journalist Ray Reece, author of *Sun Betrayed*, a book about solar energy and how it has been consistently undercut by the automotive and petroleum industries; Bob Russell, a former University of Texas literature professor turned housing cooperative manager; writer T. Paul Robbins, and others.

The two groups eventually began meeting together in what they dubbed the Tuesday Evening Lemonade Club. Its members did far more than sip sweet beverages and talk. They helped spark a substantial energy-efficiency and solar industry in Austin, convinced the city to study and eventually adopt major energy-efficiency programs, and ultimately convinced voters to change their minds on South Texas 1 and 2.

"The Lemonade Club started out as just people talking about what interested them. It eventually became kind of an incubator for new faces in the city's political scene," says Robbins.

The Club's first overtly political act was lobbying City Council member Lee Cooke to establish a new citizen board on renewable resources. Cooke, a conservative businessperson and nuclear power proponent, agreed: "I was inspired by the intellectual debate of why it made sense to look at renewable energy and conservation. If you are going to diversify your energy base, you need to explore all the options and not sell any short."

In 1978, the City Council formed the nine-member Renewable Energy Resources Commission, with Ray Reece as its first chair. Others on this purely advisory panel included journalists, architects, and energy activists. A few had extensive experience in conservation and renewable energy, but most had only the enthusiasm to learn about them. Soon afterward, a companion board called the Energy Conservation Commission was formed. The City Council also committed public funds to "alternative" energy for the first time when it created the Office of Energy Conservation and Renewable Resources. With a full-time staff of two, this office coordinated the advisory panels and turned some of their recommendations into action.

All this was basically behind-the-scenes work, invisible to the average Austin resident. Though important from an organizing perspective, the effort might never have flourished without broad citizen support. Several efforts sowed and nurtured this support. T. Paul Robbins organized a Solar Speakers Bureau to spread the renewable energy message. Representatives made solar presentations to Austin schoolchildren, spoke before community organizations, the PTA, neighborhood groups, and the Chamber of Commerce. They appeared on radio and television, and held workshops for local writers and journalists. Other folks took a direct action approach, papering Austin with thousands of "No Nukes" and "Go Solar" bumperstickers, and devising creative, visible protests that got lots of media attention. These efforts introduced solar power to thousands of Austin residents and

*In addition to two photo-voltaic power plants, the City of Austin Electric Utility Department uses small-scale PV applications, such as these PV panels mounted on top of ECHO Village's community center. The Elderly Cottage Housing Organization (ECHO) helps low-income seniors save money on their utility bills.*
Photo credit: City of Austin Electric Utility Department.

businesspeople, including many community leaders, and continued to cast doubt on the South Texas nuclear plant.

The Austin Women's Appropriate Technology Collective also spread the word. Founded in the mid-1970s, the collective taught women basic carpentry skills, solar water heater installation, weatherization, and more. Collective members worked with students from a local public school to build a solar greenhouse, and taught continuing education courses in passive solar design and weatherization.

For immediate, heartfelt community support, nothing topped Sun Day in May of 1978. This four-day festival, sponsored by the Austin Solar Energy Association with help from Tuesday Evening Lemonade Club members, featured solar power demonstrations, recycling and energy conservation booths, and lots of food and music, Austin-style. Thousands of people who would never have attended a lecture on alternative energy stopped by the festival and came away with a new appreciation for solar power. Lemonade Club members still point to Sun Day as a turning point for Austin. It gave renewable energy strong popular support, where before its backers were mostly hard-core energy activists.

The mood in Austin in 1978 mirrored the national uncertainty about the country's energy future. People recalled gasoline lines, and reports of problems at nuclear power plants regularly appeared in the newspaper and on television news. Fledgling solar and wind industries were trying to get off the ground, supported by generous research and development funding from the Carter administration and a host of renewable energy tax credits. "Energy dominated the politics of the day; we were no different in Austin," says Robbins.

With backing from area residents, the Tuesday Evening Lemonade Club managed to put a new energy agenda on the table by late 1978. What's more, renewable energy supporters had been appointed to key advisory positions on the Renewable Energy Resources Commission, the Energy Conservation Commission, and the Electric Utility Commission. The City Council, aware of blossoming interest in alternative energy and dissatisfaction with the troubled nuclear power plant, scheduled a second vote on Austin's share for early April of 1979.

That simple vote turned into what pundits called a national referendum on nuclear power. Reporters from the country's major newspapers and television stations, a crew from "60 Minutes," and even journalists from Germany and Japan covered the election. Why all the attention? Just ten days before, a faulty valve at the Three Mile Island nuclear power plant in Pennsylvania had allowed radioactive cooling water to escape. The malfunction caused a partial meltdown of the reactor core, released some radioactive steam into the atmosphere, and triggered a national re-evaluation of nuclear power. In Austin, what had initially been an intense but relatively intellectual debate over the merits and problems of nuclear power turned highly emotional. One example: The week before the vote, then-mayor Carol McClellan appeared in television commercials with her family. She said she loved her children, would never do anything to harm them, and believed that nuclear power was a safe, healthy way to make electricity.

Austin residents opted to stick with nuclear power by a slim margin, 53 percent to 47 percent. The mayor's appeal undoubtedly influenced some undecided voters. So did ongoing reports of the situation at Three Mile Island. The week of the election coincided with efforts by many in the nuclear industry to downplay the public's fears. News reports during this period emphasized successful containment of the reactor core and cooling water rather than potential problems at the plant. Some observers believe that had the vote been taken closer to the day of the accident, when fears over radioactive releases were more immediate, or later, after the full extent of the damage had been revealed, Austinites would have jettisoned their shares in South Texas 1 and 2.

Just as important, a significant number of voters weren't sure that Austin could meet the demand for electricity without its 400-megawatt share of the nuclear power plant's output. Unofficial polls taken around that time indicate that many residents believed they had no real alternatives to South Texas 1 and 2. But there *were* clear alternatives, and renewable energy advocates eventually defined and explained them in two crucial studies.

The Office of Energy Conservation and Renewable Resources published its "Comprehensive Community Energy Management Plan" in 1981. Prepared by a 135-person panel that included many citizen volunteers, the plan included an audit of Austin's energy resources and calculated its annual consumption—the equivalent of a fully loaded coal train stretching from El Paso to Austin, roughly 500 miles long. The report made scores of recommendations to maximize energy savings, minimize fossil fuel use, and benefit all Austin residents and businesses. These ranged from containing urban sprawl as a way to use less transportation fuel, to promoting residential weatherization programs and increasing the use of solar energy.

Also in 1981, the Renewable Energy Resources Commission published its "Renewable Energy Development Plan," known as the

*Austin community activists worked to elect and re-elect Tuesday Evening Lemonade Club member Roger Duncan to a city council seat. Running on a platform promoting renewables and conservation in the city's energy mix, Duncan's victory catapulted activists into city-wide decision-making roles. Duncan, kneeling in the second row with campaign manager Mary Ann Neely behind him, savors the victory with Peck Young (front) and T. Paul Robbins (center rear).*

Photo credit: T. Paul Robbins.

RED Plan. Researched and written by eighty volunteers, the plan outlined a way for Austin to achieve "energy independence through maximum reliance on alternative energy technologies and renewable energy resources." Its recommendations included:

- New building codes requiring contractors to include renewable energy technology in new buildings and to install energy-efficient heating and cooling systems;
- Zoning laws that encouraged builders to orient homes and other buildings for maximum sun exposure during the winter, and minimum exposure during the summer;
- Requiring the city-owned utility to investigate and adopt technology for generating electricity from renewable energy sources;
- Developing an efficient mass-transit system, since transportation accounted for more than one-third of the city's energy consumption.

In 1981, Austin's fermenting energy awareness bubbled over into the City Council election. Tuesday Evening Lemonade Club members Roger Duncan and Larry Deuser both ran for the council. (Deuser, an electrical engineer who worked in the defense industry, was appointed to the Electric Utility Commission in 1976 as a pro-nuclear voice. He later changed his mind when he realized nuclear power would cost

Austin far more than anticipated.) Both candidates made energy into a major campaign issue. Both won by wide margins. Their election meant that two of seven city councillors strongly favored including renewable energy and conservation in the city's energy mix. Three others on that council generally sided with Duncan and Deuser on energy issues.

This election marked another major turning point in Austin's energy story. Solar energy and conservation proponents moved from being outsider activists with purely advisory roles to insiders with crucial decision-making powers. For the next several years they kept energy issues on the front burner and built an official structure for carrying out their plans. Today, the Energy Services Division of Austin's Environmental and Conservation Services Department (ECSD) employs more than sixty workers, and handles a yearly budget of $11 million.

From 1981 to 1987, energy conservation and renewable energy were basically "mom and apple pie issues," says Duncan, who is now the assistant director of the ECSD. During that period, Austin began a number of programs which are today firmly embedded in the city's structure.

**CONSERVATION POWER PLANT**      Instead of building new fossil-fuel–fired generating plants to meet projected growth in electricity demand, the Mayor and City Council took a different tack. In 1983 they approved a Conservation Power Plant (CPP)—based largely on the RED Plan and the Comprehensive Community Energy Management Plan—that was designed to save 553 megawatts between 1983 and 1997. One report described the plant this way:

> Numerous programs and hundreds of thousands of individual actions replace the boilers and turbines usually associated with electrical generation. Energy conservation opportunities, such as increasing the efficiency of air-conditioning equipment, provide the "fuel" for the programs. In other words, the conservation power plant is spread throughout the city, in backyards and attics, on the unshaded roofs of buildings, and on the drawing boards of builders and developers.

Like the grain of sand that triggers pearl formation, the CPP crystallized Austin's energy future. It committed the city to a host of innovative energy programs, many aimed at reducing energy consumption or making certain that consumption was as efficient as possible. Other programs spawned during this time substituted renewable resources such as the sun, water, and waste for fossil fuels. Without the CPP, energy managers estimated that Austin would need 2,700 megawatts of generating capacity by 1996; with it, only 2,100 megawatts.

Although the goals were a bit ambitious, the basic idea was a success. From 1983 to 1992, the programs created by the CPP saved Austin 190 megawatts at a cost of roughly $350 per megawatt. That's about half the cost of building new fossil-fuel–fired electricity generating plants, which run approximately $700 per megawatt for the most efficient gas-fired alternatives available today. Moreover, these saved megawatts actually reduce pollution by displacing fossil fuel use.

The creators of the CPP had no crystal ball. Their vision of what Austin would be like in 1992 didn't quite match the reality of Austin in 1992. City growth never hit the projected 5 percent mark, and natural-gas prices fell rather than rose. Some parts of the original CPP became unrealistic, such as a planned manure-to-energy power plant. As part of a master plan for the 1992 City Council, Roger Duncan and his associates at the Environmental and Conservation Services Department reviewed every aspect of the original Conservation Power Plant. They found that some of its goals and technology were out of date, an understandable conclusion given the pace of innovation in the energy field. The City Council scrapped the CPP and instead set goals for minimum electricity savings of 21 megawatts a year for the next ten years; now the council is considering doubling this baseline goal and removing an existing power plant. The new energy plan leaves renewable energy development to the Electric Utility Department.

**CONSERVATION EFFORTS**    To date, Austin's energy conservation efforts have contributed the bulk of the Conservation Power Plant's "negawatts," to use a term for conserved electricity coined by efficiency guru Amory Lovins. These varied efforts coalesced into an extremely comprehensive program, offering financial incentives and education programs for every end use: existing residences, new homes, apartments, low-income dwellings, commercial buildings, and municipal facilities.

These conservation programs were administered by a newly created city department quite distinct and separate from the city utility—the Environmental and Conservation Services Department (ECSD). Ultimately the utility took charge of alternative energy development on the grid, including solar power plants and peak load control programs, and ECSD took charge of conservation in homes and commercial buildings. This division was based on the philosophy that conservation measures would be more aggressively pursued by an independent, consumer-oriented agency unhindered by potential conflicts of interest.

**Appliance Efficiency Program.** Air-conditioning gobbles up vast amounts of electricity in Austin, roughly 35 percent of the electricity produced each year by the municipal utility. In the early 1980s, energy-efficient air conditioners were beginning to appear on the market, though they cost much more than run-of-the-mill models. At the time, the average air conditioner's Energy Efficiency Rating was about 7.5. (This number is calculated by dividing the cooling rate in BTUs by the electricity input in watts. The higher the EER, the more cooling per unit of electricity.) The Appliance Efficiency Program offered buyers $500 rebates on more efficient air conditioners that carried a 9.0 rating, making them the least expensive to buy. "After a while, contractors just couldn't sell the 7.5-rated models, so they stopped stocking them," says Bob Russell, a former Tuesday Evening Lemonade Club member who now works in the communications office of the city's ECSD.

As demand shifted from cheap, inefficient units to mid-priced, efficient ones, the city changed the building code, requiring that air conditioners in new buildings be rated 9.0 or higher. After that, rebates were only offered on even more efficient machines rated 10 or 12. "First you lead with market incentives, then bring the building code up behind it. That creates a floor so you can't go backward in tough economic times," says Russell.

The Appliance Efficiency Program still distributes rebates for air conditioners, heat pumps, solar water heaters, heat pump water heaters, and heat recovery systems. Through the end of 1991, the city estimated this program had saved almost 80 megawatts of power.

**Home Energy Audits/Weatherization.** Few of Austin's homes or businesses were built with energy use in mind. Plans for the Conservation Power Plant suggested that simply retrofitting existing homes and apartments could save 150 megawatts. The city began offering free audits to measure a home's energy use and to pinpoint ways to make it more efficient. Elderly and low-income homeowners are eligible for free weatherization work, including attic insulation, weatherstripping doors and windows, wrapping water heaters, installing solar screens, even reglazing windows. Other homeowners can apply for zero-interest loans for weatherization and energy-efficiency projects.

**Green Builder Program.** This comprehensive program was one of just twelve local initiatives—the only one from the United States—honored at the United Nations Earth Summit in 1992. A brochure describing the program says: "The program's goal is to shift residential building practices toward sustainable approaches that conserve not only energy, water, and other natural resources, but also preserve our environment, strengthen our local economy, and promote a quality of life that is more rewarding to the citizens of Austin." In a nutshell, the Green Builder Program rates a new home using several criteria, such as: how efficiently it uses water and energy; the materials used to build it; whether building materials are recycled during home construction; and whether recycling has been built into the design with a kitchen recycling center or backyard composting system. Four stars identify homes with "superior" sustainability ratings, one star those that are merely "good." Austin-area buyers are beginning to use the ratings to compare new home styles and prices, and program coordinator Laurence Doxsey says the city is lobbying to have all new city housing meet at least a two-star rating.

**THE APARTMENT PROGRAM**   Conservation in apartments is rare. Since energy savings are realized in lower tenant utility bills, landlords have little incentive to invest in conservation. Conversely, tenants usually do not consider or cannot afford to make expensive retrofits to a building they don't own. Austin's program, one of the few in the country, offers landlords generous rebates to install efficient air-conditioning or heat pumps and to make extensive retrofits during remodeling. Owners can also use the new conservation features in sales promotions, making the apartments easier to rent. Approximately three-thousand apartments have been improved by the program, resulting in lowered energy bills in 30 percent of the multifamily dwellings in Austin.

**BOND ISSUES**   Like any major municipal program, the Conservation Power Plant required a healthy dose of capital. In 1983, voters approved a $39 million municipal bond for conservation and renewable energy programs. This included $10 million for a hydropower plant (not yet built) on the Colorado River, which runs through the center of Austin, and $29 million for other renewable energy projects. Unlike earlier energy votes, this one sailed through easily, possibly because voters were beginning to see the value of conservation in their own utility bills.

Austinites also backed a second bond issue in 1986, this one for $60 million to fund the entire Conservation Power Plant. Unfortunately, the state Attorney General ruled this bond was unconstitutional. Texas law forbids municipalities from borrowing money and loaning it to private individuals, precisely what the low-interest and no-interest weatherization loans would do.

**RENEWABLE ENERGY USE**    Though non-fossil, non-nuclear energy still contributes only a tiny fraction of Austin's annual energy supply, the city must be considered one of the country's leaders in promoting renewable energy. Its work with photovoltaic generating systems in particular is helping map this technology's path to commercialization.

To meet the Conservation Power Plant's aggressive goal of 200 megawatts or more of renewably generated electricity, the municipal utility hired a young specialist in renewable energy systems fresh out of graduate school. John Hoffner's task, directing the utility's program on alternative energy technologies, wasn't an easy one. Like many organizations, the City of Austin Electric Utility Department resists change, especially the fairly radical change of relying on photovoltaic generators.

To drum up support for his first major project, a 300-kilowatt photovoltaic power plant, Hoffner took some of the utility's engineers, designers, and maintenance and power production workers out to Sacramento, California. There they toured the world's largest photovoltaic plant, a 2-megawatt model built right next to the now-closed Rancho Seco nuclear power plant. "They saw it was a real generating plant with transformers, switches, transmission lines, and everything," Hoffner recalls with a laugh. He also sponsored brown-bag lunches to introduce workers in all departments to alternative energy technologies. With strong support from the top, he has helped the utility come to see photovoltaics as a reliable and often cost-effective alternative to extending transmission lines.

"Solar-generated electricity is a perfect match for Austin," explains Hoffner, now the utility's project manager for alternative energy. Peak times for electricity use in the city occur on hot

*John Hoffner, the Austin utility's alternative energy manager, believes solar-generated electricity is a perfect match for Austin's energy use patterns; peak air-conditioning days are also the best time for PV. Hoffner stays on top of the utility's various PV projects; this photo was taken during a tour he led at the AYH Hostel, where PV panels now generate electricity for the 50-bed hostel.*
Photo credit: Kevin Gallagher.

summer afternoons, when residents and businesses have their air conditioners going full blast.

The showpieces of Austin's venture into renewable energy are two fully functional photovoltaic power plants. Both can generate up to 300 kilowatts of electricity, though they use different methods. The city built its PV300 plant in 1986 for approximately $3 million. It was never meant to make a dent in generating capacity, but rather to give utility workers experience with this new technology and prove its safety and reliability. Across town at the 3M Austin Center, another photovoltaic plant generates a maximum of 300 kilowatts from atop the building's parking deck.

The two plants are teaching local energy managers and national experts valuable lessons about photovoltaic electricity generation. The PV300 plant uses flat-plate solar cells that passively track the sun as it moves across the sky, while the 3M plant uses special Fresnel lenses to concentrate sunlight onto strips of photovoltaic cells. PV300 uses 2,620 square meters of solar collectors that convert the sun's energy into electricity with a 14.5 percent efficiency. The concentrator, in comparison, uses 2,006 square meters of lens area and only several hundred square meters of solar cells, which convert sunlight into electricity at an efficiency of 17.5 percent. On paper, the concentrating technique appears to be the clear winner, since it requires fewer square feet of expensive PV modules. In practice, in Austin, it isn't. The concentrator technology requires direct light shining on the lenses and works exceptionally well in dry, desert environments. But Austin's high humidity and haze scatter sunlight before it reaches the ground, so the concentrator system averages only about 270,000 kilowatt-hours of electricity each year. The flat-plate collectors, on the other hand, handle diffuse sunlight much better and churn out 440,000 kilowatt-hours a year. There's a crucial lesson here that applies to every alternative energy project: the technology must be carefully matched with local climate, geography, and land conditions.

Each of the two plants cost approximately $10,000 per kilowatt of generating power, well above the $1,000-per-kilowatt price tag for a gas turbine generator. And they add only a minuscule amount to the city's generating capacity. Yet their contribution has been invaluable. "All this testing is preparing us for the day when the cost of PV comes down and it is time to put plants like this on line. We will be ready," says Hoffner.

The city is also investigating a number of small-scale photovoltaic applications, some paid for with federal oil overcharge funds. In fact, these so-called niche applications may be ready for mainstream commercial use long before central PV power plants (see Chapter 3). For example, rooftop photovoltaic panels now generate electricity for Austin's fifty-bed youth hostel. During the day, when demand is low, excess electricity is sold back to the utility; at night, the hostel draws electricity from the grid. On most days, power generation from the PV system equals or even exceeds consumption for a typical Austin home, and would bring electric bills down near zero for many months in the year (a practice known as *net metering*). To test the fit between photovoltaics and multi-family housing, the utility installed solar panels at a small housing development for elderly, low-income residents. And the city's new convention center has been fitted with a bank of PV panels that contribute about 3 percent of the building's electricity demand. Austin even uses sunlight to power the blinking arrows that guide traffic around construction sites.

*After three-and-a-half years with no complaints, Alicia Lee (left) and Ethel Deshay (right), residents of ECHO Village, a subsidized elderly housing project, perceive photovoltaics not as a novelty but as an ordinary part of their lives.*
Photo credit: National Renewable Energy Laboratory (NREL).

Working photovoltaics into the energy mix takes careful coordination between engineers, designers, and power production workers. "In most cities, solar power isn't an option because there's no sheet in the design manual to suggest that photovoltaics might be a possibility," says Hoffner. Austin's design manual does make reference to photovoltaics. Hoffner and his team are creating complete design and evaluation tools to help their utility and others around the country use photovoltaics more easily. These will allow utility planners to point customers to standard, off-the-shelf equipment that will provide them with the electricity they need.

Solar energy isn't the only renewable resource that the city plans to develop. In 1986 it received a license to build a 3-megawatt hydroelectric power plant at the Longhorn Dam on the Colorado River, which runs through Austin. Falling natural-gas prices and excess generating capacity, however, have contributed to delays in construction. But the municipal wastewater treatment station is up and running, collecting enough methane gas from decomposing sewage to heat the plant's boilers and generate up to 800 kilowatts of electricity.

**UTILITY RATES CHANGED**     Like most utilities in the 1970s, Austin's municipal utility calculated rates on a bigger-is-better scale: those who used the least amount of electricity paid the highest per-kilowatt rate, and the largest users paid the lowest rate. While attractive to business and industry, this pricing scheme contradicted the city's developing philosophy in energy management. Peck Young puts it this way: "Why charge rates that encourage the waste of electricity by corporations and large users, yet penalize individuals to the point that some people weren't putting out Christmas lights?"

In 1981, after five years of behind-the-scenes negotiations by the Electric Utility Commission, voters approved a proposal that dramatically changed electric rates and electricity use. The new scale charged homeowners and small businesses a flat rate for 500 kilowatt-hours a month. Anything over that amount was billed on an inverted scale, with rates climbing along with consumption. Giant industrial customers were charged a flat per-kilowatt rate. Roughly three-quarters of Austin-area homeowners and small-business owners saw smaller electric bills as soon as the new rate structure was put in place.

**NUCLEAR POWER SHARE DROPPED**     Not long after the 1981 election, the new pro-renewables City Council once again asked voters if Austin should continue to hold shares in the partially built South Texas 1 and 2. This anti-nuclear campaign was orchestrated by Peck Young and Mary Ann Neely, who had just engineered Roger Duncan's election to the City Council. Using the same organization, they targeted only undecided voters, and drove home two key points: the two reactors would cost taxpayers far more than they ever imagined; and renewable energy could provide a reliable, safe, inexpensive alternative to nuclear-generated electricity. This simple message did the trick. A hefty majority instructed the Mayor and City Council to sell off Austin's one-sixth share of the plant. The vote may have been more symbolic than concrete, however. Unless the city can find a buyer for its share—none have come forward so far—the city will continue to be financially responsible for South Texas 1 and 2, and receive electricity from them.

**SECURING THE GAINS**     From 1981 to 1987, strong popular and political support helped Austin lay a solid foundation for its energy future. Programs that were considered too "far out" in many cities became part of Austin's official bureaucracy and infrastructure. These important strides weren't accomplished without political struggle, negotiation, and community outreach. "None of this would have worked without an extensive education program," says Bob Russell, whose work at Environmental and Conservation Services Department (ECSD) includes public information programs.

Borrowing ideas from Austin's successful activist outreach efforts, the Resources Management Department and later the ECSD sponsored school programs, gave presentations to neighborhood groups and churches, and used slide shows for talks at Chamber of Commerce and Lion's Club luncheons. In the beginning, energy education seminars were held in each section of the city in order to train people who wanted to self-weatherize their homes or who wanted to be part of the pilot retrofit programs. Brochures, pamphlets, and flyers were mailed to Austin residents, explaining the conservation efforts and rebate programs. The local public television station even produced and aired a video series on energy. As the outreach programs matured, professional advertising was produced for direct mailings and media, and exhibitions appeared at trade shows and public events. This level of public education and outreach contributed greatly to Austin's renewable energy and conservation successes.

Virtually every step of the way, Austin translated its forward-thinking energy programs and policies into regulations or laws. So, in 1987, when the flame flickered and key renewable energy supporters left the political scene, the programs they brought to life weren't snuffed out as well. In fact, most even survived the challenges of the late 1980s, when the Texas economic boom turned to bust. "We institutionalized a pretty positive attitude toward renewable energy and conservation in the 1980s," says former Chamber of Commerce president Lee Cooke. During the general belt-tightening, Austin companies laid off thousands of employees and began searching for ways to cut costs. City government was no exception, and energy rebate programs appeared to be prime targets. One City Council member blasted appliance efficiency rebates as nothing more than "socialized air-conditioning," and unsuccessfully lobbied to gut rebate and other conservation programs.

South Texas 1 and 2 began producing electricity in 1989, bringing the city's generating capacity to 1,900 megawatts. But peak demand hovered at around 1,200 megawatts and baseload needs averaged less than 600 megawatts. "It was hard to call for energy conservation when there was no real need to do it, when much of the city's generating capacity sat idle each day," says Bob Russell. So energy managers quietly shifted their conservation arguments from economic to environmental grounds, pointing out that efficiency efforts kept tons of pollutants like carbon dioxide and nitrogen and sulfur oxides out of Austin's air. They justified the cost for conservation and renewable energy programs as relatively inexpensive investments in maintaining Austin's quality of life.

> CONTACT: *John Hoffner, Project Manager, Alternative Energy, City of Austin Electric Utility, 721 Barton Springs Road, 2nd Floor, Austin, Texas 78704; 512-322-9600. Roger Duncan, Assistant Director, Environmental and Conservation Services Department, City of Austin, 206 E. 9th Street, Suite 17.102, Austin, Texas 78701; 512-499-3575.*

**LESSONS LEARNED**   Like it or not, we are bound by many written and unwritten rules. In the energy arena, virtually all of them favor conspicuous consumption, fossil fuels, and traditional ways of generating electricity. Savvy homeowners and businesspeople can break those rules with impunity, and reap the benefits of energy efficiency or renewable energy. Most people, however, haven't the time nor the inclination to become energy experts, and thus live by the established rules. Changing these rules is one important element in the struggle to expand the use of renewables around the country.

Austin did change the established rules, thanks to citizens committed to their ideas who also knew how to maneuver the system. The programs they began are now part of city building codes and are firmly entrenched in city departments. "Everything we did eventually got bureaucratized," says activist T. Paul Robbins. Institutionalizing these changes made it difficult to discard them when Austin faced an economic slump in the late 1980s.

But the Austin experience also illustrates that activists must remain ever-vigilant to the possibility of counterattack, even when programs seem well established. In fact, a backlash is often most vicious immediately after an important victory, when supporters are tired, complacent, or want to move on to something else. Roger Duncan's 1983 reelection campaign, for example, was a real squeaker; only an extraordinary effort, reactivating the coalition that had elected him previously, saved his city council seat. Had he been defeated, conservation programs would not have had sufficient time to become embedded in the city's structure or the public's consciousness. Sustained, visible grassroots support and activism is vital for success over the long haul.

Changing even the smallest energy rules will have an impact on the future of renewable energy. Like wind and solar power, these laws or regulations are "site-specific," with precise meanings in each community. Make sure you are going after those rules which will make the most difference in your community. The goal here is not change for its own sake. Replacing old rules with new ones should be aimed toward a single goal—making renewable energy the norm, the way business is done, rather than an intriguing oddity.

**SOLDIERS GROVE, WISCONSIN**    Passive Sun Drive, Sunbeam Boulevard, and Sunset Avenue aren't just street names in an overly cute suburban development. They mean business in Soldiers Grove, a small southwestern Wisconsin town that has staked its very future on solar energy. The town bills itself as America's first "Solar Village." This strategy—originated, studied, promoted, and carried out by community members—has helped Soldiers Grove recover from a series of devastating floods and the economic decline faced by small rural towns all across the country.

Every so often throughout this century, the normally docile Kickapoo River has jumped its banks and inundated Soldiers Grove's business district, which is built right on the shallow floodplain. Water and mud repeatedly damaged homes and stores and drove businesses out of town. In the mid-1970s, several town leaders began talking about moving the entire downtown out of the floodplain. Their other option was paying the Army Corps of Engineers to build an expensive levee that couldn't completely guarantee protection against future floods.

The relocation idea struck a chord with the town's six hundred or so residents. The Village Board formed a Citizens' Planning Committee and hired Tom Hirsch, an architect and community planning specialist who had recently moved to the Kickapoo Valley, as full-time "relocation coordinator." Exactly how the solar village concept became part of this plan is lost in the haze of memory. William Becker, former publisher of the town newspaper, recalls "blurting out the idea one day during an interview about a possible relocation with a Chicago radio station."

The solar village concept intrigued residents, business owners, and civic leaders alike for a host of reasons. Heating costs accounted for a large fraction of business expenses, and cost projections pointed ever upward. During the 1970s, energy supplies fluctuated wildly, and

*America's First Solar Village*

*Big snow drifts and sub-zero temperatures don't deter customers of the Solar Town Pharmacy. With the solar attic to gather heat, and high levels of insulation, customers stay warm and pharmacist Don Stabner saves money. Nearly all of the businesses in Soldiers Grove are solar heated.*

Photo credit: Nancy Cole.

memories of oil embargoes were fresh in residents' minds. The independence that solar heating represented, and the fact that it could prevent an annual flood of energy dollars from leaving the town each year, appealed to many. It also could give the town a new identity, one more in keeping with the civic leaders' visions for the future. Instead of being known as just another river town, they hoped to attract new businesses to Soldiers Grove, and possibly make the area a hub for medical services.

Relocation coordinator Tom Hirsch got citizens and planners to study the town's potential for solar heating. Residents helped measure the hours of sunlight at different locations and different times of the year. This work ultimately led to shifting the site for a new downtown, since the initial one was in the shadow of several hills to the north.

The worst flood in a century turned talk and theory into action in 1978. Rather than once again repair damaged buildings, the citizens of Soldiers Grove voted to move their downtown. They also agreed that all new buildings should get as much heat as possible from the sun. Since 1978, more than twenty solar-heated stores and businesses have been built in downtown Soldiers Grove. These include Turk's IGA grocery store, the Solar Town Pharmacy, and Community Library on Passive Sun Drive; Peoples State Bank, Olson's Mobil Service Station, and the Country Garden Restaurant on Sunbeam Boulevard; and Kickapoo Valley Medical Clinic and the Post Office on Sunset Avenue. A sixteen-unit housing complex for the elderly and several homes also rely on solar heat.

Most of the new buildings use high levels of insulation (R-36 in walls and R-72 in attics and roofs) to hold in heat, thus reducing daily demand. And most use solar attics for gathering solar energy. The 3,000-square-foot medical clinic, for example, has its south-facing roof built on a 60-degree angle. Almost all of this surface is covered with fiber-glass-reinforced polyester glazing. Light streaming through this glazing heats the air in a highly insulated attic; blowers send this air underneath the slab floor, where it warms a series of concrete blocks and gravel channels. The glazing treatment not only stores heat well into the night, but also ensures that it is evenly distributed. A more common approach involves merely blowing warm attic air throughout the building.

Skeptics have argued that solar heating won't work in places like Wisconsin—northern latitudes with long, cold winters. Tell that to Don Stabner, who owns the Solar Town Pharmacy. Pulling out his records for the past few years, he proudly points to heating bills that average less than $600 per year for a 2,150-square-foot building. "Just from a business perspective, I'm saving money," he says. "When things are close financially, this solar heating can make a real difference." Town officials recruited Stabner for the new downtown from North Dakota. He's pleased with his move, and now promotes the solar village through his spot on the Community Development Board. "I'm proud of what we've accomplished here," he says.

The first solar buildings were constructed in the spirit of renovation and relocation. To maintain that commitment, the townspeople decided to put their forward-thinking attitudes into law. Since 1980, section 2.06 of the city ordinances protects solar access for all new buildings from shading by other buildings or by trees or shrubs. Section 2.08 mandates that nonresidential buildings require no more than 8 BTUs per degree day,* per square foot, for heating. Furthermore, section 2.10 requires, where feasible, that new nonresidential buildings obtain at least 50 percent of their heat from the sun.

These ordinances wisely focus on performance goals rather than specifying technology or construction techniques. They give owners and builders plenty of leeway in meeting the energy targets. So far, most of the new buildings rely on solar attics, but there are also plenty of innovative—and often site-specific—variations. Turk's IGA, for example, supplements solar energy with heat recovered from its commercial

---

*Each degree day is equivalent to an outside temperature one degree cooler than 65 degrees F over a twenty-four hour period.

refrigerators and banks of fluorescent lights. In most years, this combination makes it possible for the grocery to avoid using auxiliary fuel. Like other solar buildings, the store has a back-up heat source; in this case, liquid propane. (Given their thick insulation and the sun's input, even large commercial buildings need only home-sized furnaces for backup heat.) Not only is propane readily available, but Soldiers Grove residents are looking ahead to the day the town makes its own bottled gas from sewage sludge, crop residues, or other farm waste.

While the relocated solar village has been a technical success, residents and leaders had hoped it would somehow revitalize the town and attract new industry or business. That hasn't happened. A few of the town's political leaders, desperate to attract new businesses to the town, even view the rules as liabilities. One such critic is Irving Davidson, the current mayor. In fact, the Village Board has begun granting exemptions to the solar ordinances.

Despite all evidence to the contrary, some people see the solar requirements as an obstacle to economic development. Ask any Soldiers Grove business owner, and he or she will say that solar heating pays for itself and then some. But it's hard to convince others of that simple truth. "If it was generally accepted in this country that renewable energy is a benefit and not a burden, an advantage and not a barrier, then Soldiers Grove wouldn't be fighting this battle alone," says Becker, who now works for the U.S. Department of Energy's Office of Conservation and Renewable Energy.

CONTACT: *Ardelle Knudson, Town Clerk, PO Box 121, Soldiers Grove, Wisconsin 54655; 608-624-3264.*

## *Solar Access Laws*

**PORTLAND, OREGON**    State-of-the-art solar access laws adopted by Portland, Oregon, and twenty-one surrounding communities clearly change the rules to favor renewable energy use. This three-part effort both encourages and protects the use of solar energy:

• Solar Access Standards for New Development require that 80 percent of the lots on a new development front onto streets that run east to west; they also require that these lots have a 90-foot or greater north-south exposure. This "orientation" ordinance ensures that the majority of homes will be built with unobstructed, unshaded south-facing walls, allowing them to gather the most sunlight possible. (See diagram.)

• Solar Balance Point Standards for existing homes prevent new homes or home additions from shading a neighbor's roof or walls. This law

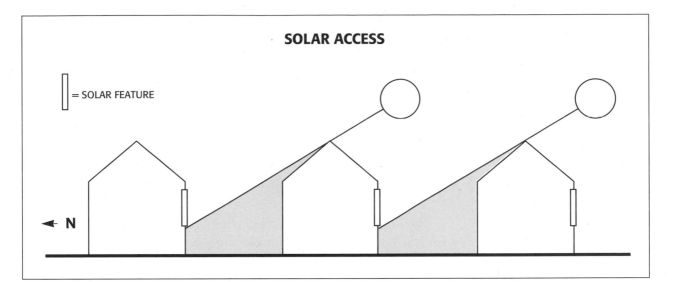

**SOLAR ACCESS**

= SOLAR FEATURE

← N

ensures a steady stream of sunlight for people who build a passive-solar home, install a solar water heater, or invest in photovoltaic panels.

- The Solar Access Permit Ordinance prohibits people from planting "solar unfriendly" trees (i.e., evergreens or deciduous trees with dense clusters of branches that block sunlight even when there are no leaves on them) in locations that would potentially shade south-facing solar collector areas. Local researchers have identified more than 250 kinds of trees that will give Portland-area residents shade in the summer but won't block sunlight in the winter.

These fairly simple laws provide a fundamental starting point for long-range solar development by ensuring that anyone who wants to use solar energy will always have access to sunlight. Making solar access part of every new development and protecting people's right to collect sunlight also infuses the concept of renewable energy into mainstream thinking. This gives renewable energy and energy conservation a higher profile, which can in turn stimulate demand for these technologies. Energy analysts all agree that low demand is one of the major barriers to renewable energy use in the United States today.

Although the Portland-area laws provide a solid model, they were developed only after several years of intense research in the communities involved. Solar and wind energy are so site-specific that any local ordinance must be based on adequate local research. What works in Portland,

*Portland's solar access laws ensure a steady stream of sunlight for people who invest in solar technologies. Planners and builders calculate available solar access by measuring unobstructed sunlight on December 21, the day of the year with the least sunlight.*
Credit: City of Portland, Solar Access Ordinances.

*To ensure adequate solar access, the solar collector should receive unobstructed sunlight.*
Credit: Planning Solar Neighborhoods, California Energy Commission.

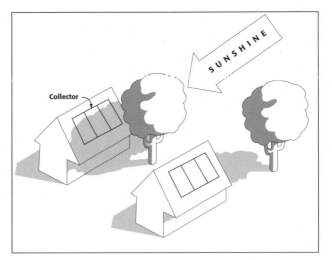

Collector

SUNSHINE

Oregon, with its modestly hot summers and cool, cloudy winters might not be right for Portland, Maine, with its modestly hot summers and cold, snowy winters.

CONTACT: *Susan Anderson, City of Portland Energy Office, Portland Building, 1120 SW 5th Avenue, Portland, Oregon 97204; 503-823-7222.*

## Comprehensive Solar Access Program

**SAN JOSE, CALIFORNIA**  Rather than using laws to lower barriers against renewable energy use, San Jose, California, relies on a set of "solar access" guidelines to assure unshaded sunshine availability for new home construction and market incentives for builders that favor renewables. This city-organized and city-run program illustrates an important lesson about guidelines: just creating or publishing them isn't enough—making them work requires continued support, education, and outreach.

As part of its Environmental Services Department, San Jose established a Solar Access Program with a three-part mission: designing easy-to-use techniques for determining minimum available unshaded sunshine in winter; teaching builders that planning developments with a maximum of south-facing homes would cost no more than conventional development; and informing both homeowners and builders of the economic and aesthetic benefits of unrestricted access to sunshine for heating, energy, and interior lighting.

Program staffers work closely with architects, developers, builders, and homeowners. A recently published "Solar Access Design Manual" provides practical design tips for single- and multi-family homes that help builders and designers make unrestricted availability to sunshine an integral part of their work. The manual also presents three case studies. Each takes a conventional development plan and reorganizes it to vastly improve the favorable orientation to the south while maintaining density and lot sizes. Hands-on professional workshops clarify and reinforce the manual's lessons.

Non-binding guidelines rarely work, especially when they suggest actions that businesspeople see as economically unsound. On the surface, designing for solar access doesn't appear to reward a designer or builder for the extra work. But research coming out of the Solar Access Program shows that homes designed with solar energy

---

**STATES WITH SOLAR ACCESS LEGISLATION**

As of early 1994, states with solar access legislation include: Arizona, California, Colorado, Connecticut, Florida, Georgia, Idaho, Kansas, Kentucky, Minnesota, New Jersey, New Mexico, North Carolina, North Dakota, Oregon, and Virginia.

benefits in mind save money for both owners and builders. Orienting the windows to the south saves homeowners up to 16 percent in energy costs annually, and buyers place a high value on sunny homes, energy efficiency, and comfort. Thus, homes designed with solar access in mind should be more attractive to buyers than conventional homes. Solar Access Program research also shows that simply orienting buildings toward the sun can save builders substantial sums of money—by reducing total energy requirements, other energy conservation measures can be down-sized or even eliminated while still meeting energy-efficiency standards.

Finally, San Jose provides a very practical incentive for developers, designers, and builders who follow the guidelines. The city's Planning Department looks favorably on plans that incorporate the recommendations on solar access and landscaping, and tends to speed them through the planning process.

CONTACT: *Mary Tucker, City of San Jose, Environmental Services Department, 777 N. First Street, Room 450, San Jose, California 95112; 408-277-5533.*

**LINCOLN, NEBRASKA** Lincoln, Nebraska's Community Unit Plan encourages "creative design in buildings, open space, and their interrelationship, while protecting the health, safety, and general welfare of existing and future residents of surrounding neighborhoods." One section of this plan promotes renewable energy use by allowing builders who design solar-friendly residential developments to increase the density of new homes by up to 20 percent. In Lincoln, solar-friendly means laying out most streets along an east-west axis, orienting lots to ensure the greatest south-facing exposure, and building houses that won't be shaded by trees or other buildings. Other considerations include:

- Designing buildings that maximize solar energy use;
- Landscaping to reduce heating and cooling needs, such as planting windbreaks where appropriate or planting only those trees that offer good shade in the summer but don't block the sun in winter;
- Outfitting east- and west-facing windows with adjustable solar screens.

CONTACT: *Rick Houck, Planner, Lincoln-Lancaster County Planning Department, 555 S. 10th Street, Lincoln, Nebraska 68508; 402-441-7491.*

## Incentives for Developers

*Orientation of lots and houses to ensure the greatest south-facing exposure can vary up to 15 degrees east or west without significantly affecting solar gain.*
Credit: *Solar Homes for Virginia*, Virginia Office of Emergency and Energy Services.

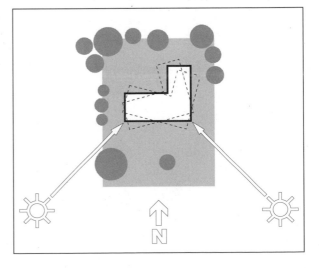

## *Energy Coordinator*

**IOWA CITY, IOWA**     Even a modest-sized city like Iowa City, Iowa (population approximately 50,000), must manage a fair amount of real estate—a school, fire and police stations, city hall, a library. Until 1978, each city department set its own policy on energy use; since no department heads were energy experts, buildings often used energy inefficiently. So in 1978, Iowa City hired its first Energy Coordinator, a move that has slashed energy use in its thirteen municipal buildings 43 percent since 1983, and saved the town more than $2 million.

The part-time Coordinator identifies potential conservation measures in city buildings and helps administer a revolving Energy Savings Loan Fund. This fund, available only to city departments and divisions, provides money for conservation projects; the loans are repaid out of energy cost savings.

The recent renovation of the Iowa City animal shelter, which for years was little more than a concrete block box, is a good example of what Energy Coordinator James Schoenfelder (who also works as the city architect) does. He proposed adding a new lobby with a wall of south-facing windows to gather sunlight for heat, a heat storage system to keep floors warm, a new heat-absorbing Trombe wall laid atop the old concrete block, and superinsulation everywhere. The expanded shelter uses 57 percent less energy per square foot than the original.

Other energy-saving ideas that Schoenfelder has put into practice include:

- Installing motion sensors in city offices that automatically turn off lights in unoccupied offices, saving up to 30 percent on lighting costs;
- Replacing incandescent lights in the community recreation center with fluorescent bulbs; and
- Covering the town pool at night, saving about $4,000 per year in heating costs

According to Schoenfelder, any city or town that spends at least $175,000 a year on energy bills can't afford *not* to hire at least a part-time energy conservation expert.

CONTACT: *James Schoenfelder, City of Iowa City, City Architect/Energy Coordinator, 410 E. Washington Street, Iowa City, Iowa 52240-1826; 319-356-5044.*

---

## SOLAR ZONING CODES ALBUQUERQUE, NEW MEXICO

Zoning codes offer another avenue for changing the rules. In its residential zoning laws, Albuquerque, New Mexico, places height limits on new buildings to ensure they do not block the sun's rays from existing buildings relying on passive solar energy, situated to the north. The code uses a solar sliding scale to calculate acceptable heights. The further a new building or addition is from the lot line to the south, the taller it can be. Buildings less than 5 feet from the line can't exceed 10 feet in height; those 35 feet or more from the line can go as high as 26 feet, the maximum allowed in Albuquerque's residential areas.

CONTACT: *Jack Bayse, Zoning Enforcement Inspector, City of Albuquerque Zoning, 600 2nd Street, NW, Albuquerque, New Mexico 87102; 505-764-1664.*

**FORT COLLINS, COLORADO**    Although improving renewable energy use one rule-change at a time definitely helps, Fort Collins, Colorado's strategic planning approach may work even better. This public/private venture started out as a survey of shading problems in the area, then developed into a forecast of local renewable energy use. The planning board ultimately selected a six-part strategy aimed at preventing development from literally overshadowing active and passive solar energy use. This strategy included:

- Adding solar energy policies to the city's Comprehensive Plan;
- Removing potential barriers to solar energy use from zoning and subdivision regulations;
- Providing city help for planting solar-friendly trees along streets, for evaluating development projects that might shade potential solar collector sites, and for reserving airspace that protects solar access to solar collectors, greenhouses, and other investments;
- Creating an office to provide residents with information on solar access and solar easements;
- Establishing a system for notifying sun-using building owners of proposed construction that could shade existing systems so that an owner can negotiate changes with the builder; and
- Including solar access planning criteria in the city's overall planning and development regulations.

The city's Strategic Plan for Solar Access, drawn up over a two-year period, involved citizens, solar energy activists, city planners, and state and local energy officials. This route was chosen because it "explicitly considers resource availability . . . The strategic planning process produces better coordination within the community, by enabling various groups to act on the same information and conclusions."

CONTACT: *Joe Frank, Project Manager, City of Fort Collins Planning Department, PO Box 580, Fort Collins, Colorado 80522; 303-221-6500.*

**CAPE COD, MASSACHUSETTS**    Cape Cod conjures up images of windy, sunswept beaches, splendid summer homes, and Yankee self-sufficiency. For the area's full-time inhabitants, though, life on the Cape means wages 25 percent below the average salary in Massachusetts, gas and electricity prices that are in the nation's top ten, and the rapid development of strip malls and tourist attractions. In an effort to balance development and preservation, growth and sustainability,

Cape Cod residents are developing a regional management plan. This effort includes an energy management plan. Since September of 1991 more than one hundred people have studied and debated several key issues: energy-efficiency incentives; the role of renewable energy technologies, such as solar thermal systems, photovoltaics, and wind and wave power; regulatory options; the role of utilities; a home energy rating system, and a cost-effective alternative to automobiles.

Since the scope of the project is so encompassing, progress has been relatively slow. The organizers are including representatives from each town or city in Barnstable County—businesspeople, residents, utility executives, and energy activists. Along the way they have uncovered some chilling statistics about this area of the Cape. In 1990, for example, Barnstable County used thirty-seven trillion BTUs of energy, costing $434 million. If environmental costs are figured in, the total soars to $654 million. Since virtually all of this energy came from imported sources, approximately $370 million left the local economy, or about $2,325 per person. While Cape Cod residents used 11 percent less energy per capita than the average Massachusetts resident, they paid 40 percent more for it.

Recommendations in the final energy plan include: setting efficiency goals for new home construction, encouraging banks to back energy-efficient mortgages, promoting solar water heating and passive solar construction, evaluating the area's potential for wind-generated electricity, improving public transportation, and choosing liquified natural gas as an appropriate alternative fuel for area vehicles.

CONTACT: *Matt Patrick, Cape and Islands Self-Reliance Corporation, 24 Collins Road, Waquoit, Massachusetts 02536; 508-457-7679.*

## *Comprehensive Energy Plan*

**SPRINGFIELD, ILLINOIS**    Changing the rules usually means changing minds. It's not that people oppose energy conservation or renewable energy use; they just need to understand the issues and economics. One way to accomplish this is with careful organization and serious coalition-building.

Springfield, the capital of Illinois, is a relatively conservative city with access to cheap energy from coal-fired plants. Yet in the 1980s, the Springfield Energy Project focused attention on energy conservation and helped residents and businesspeople look at energy use in a new way. "The Project changed the city permanently. It is now a very hip town in terms of energy issues," says Alexander Casella, Project Director.

The Springfield Energy Project grew out of a class Casella taught at Sangamon State University in which students calculated the city's energy flows. They found that Springfield paid approximately $100 million for energy in 1977, and $150 million in 1980, and that most of this money left the local economy. They also determined that energy efficiency and renewable energy could plug 40 percent of this drain.

Spurred by these numbers, Casella and other organizers created the Springfield Energy Project. They identified key areas of energy use —residential, government, transportation, local food production, commercial/industrial, waste utilization, and institutions—and created a task force for each. They invited two hundred people from a wide range of backgrounds to work on the project. Participants included housing activists, builders, bankers, scrap dealers, utility officials, physicians, newspaper editors, and many others. "It was remarkable to see people who were traditional adversaries developing a rapport with each other and really working together," said Casella. The result of their efforts, a report titled "Community Energy Self-Reliance: A Citizens' Plan for Springfield, Illinois," included recommendations from each task force. The report was adopted as a resolution by the City Council. So far, approximately 30 percent of the recommendations have been put into action.

Casella has these suggestions to offer others organizing similar rule-changing projects:

- Involve all interest groups from the start. Don't leave out those you assume will be hostile to the work. This strategy generally gets all the issues out on the table, and can often reveal surprising sources of support.
- Phrase the problem in a way that makes sense to people, and that doesn't reflect a particular ideology. The Springfield organizers calculated that the money spent annually on energy, almost all of which was paid to out-of-town businesses, was like losing one of the city's major employers, each year.
- Be confident in your data. Talk over the figures with local utility officials and other energy experts, so debates don't get bogged down in arguments about numbers.
- Assure the members of the project they will be heard, and that their participation is vital. The Springfield group promised that they would publish every recommendation passed by a majority of members on a task force.

CONTACT: *Alexander Casella, Dean, School of Public Affairs, Sangamon State University, Springfield, Illinois 62794-9243; 217-786-6523.*

## Energy Town Meeting

**ASPEN, COLORADO**     While home energy consumption fell across the country throughout the 1980s, changes in the economy and social composition of Aspen, Colorado, and surrounding Pitkin County caused energy use to rise much faster than the population. To counter this alarming trend, area residents organized the City-County Energy 2000 Committee; this volunteer committee in turn borrowed a page from traditional New England government and initiated annual Energy Town Meetings. The first Energy 2000 Forum and town meeting in October of 1991 led to the organization of citizen-staffed task forces working on a draft energy plan. The nonprofit Roaring Fork Energy Center also played a key role in compiling and analyzing the "Community Energy Action Plan." The Rocky Mountain Institute, an internationally recognized center for energy efficiency founded by Amory Lovins, helped organize the task forces and worked with the local media to publicize their work. Other key players included Aspen Skiing Company—the area's largest employer—and the Chamber of Commerce.

The energy plan, presented at the second Energy Town Meeting in October of 1992, included a comprehensive review of local energy use. The plan suggested a $1.5 million investment in energy efficiency measures, which would in turn generate more than $5.4 million in economic benefits. It called for a 66 percent reduction of non-renewable energy over the next eight years through energy efficiency and attention to renewable energy technologies. In addition, it suggested "feebates" (fees to inefficient users that would pay for rebates to those who embraced energy efficiency), and other creative ways to help home and business owners finance energy-efficient retrofits or new construction, and invest in renewable energy technologies.

More than two hundred people participated in the 1992 meeting, which also included booths and demonstrations by local suppliers of energy-saving and renewable technologies, and by designers of energy-wise buildings. The 1993 meeting followed the local Efficiency Grand Prix, a race open only to "environmentally and financially sustainable alternatives to the single occupant vehicle," such as buses, bicycles, skates, horses. The goal of future meetings is to ultimately help Aspen break its dependence on electricity from Nebraska coal-fired generating plants and become energy self-sufficient.

Any city or town could profit from an Energy Town Meeting. Aspen's was begun by former Mayor Bill Stirling, who realized that citizen involvement would be the most important component of any local or regional energy plan. "The process is as important as the end product," says Stirling. "Aspen's plan reaches out to a very diverse group in the community and works from the bottom up." While a direct connection with organized city government isn't crucial, it helps give decisions the weight of policy.

In August of 1994, the Community Office for Resource Efficiency (CORE) opened its doors. The Energy 2000 Committee pulled together an exciting funding collaborative including electric and natural gas suppliers, local government, and private enterprises. The new office has the potential to institutionalize energy efficiency and conservation innovations in Pitkin County—and the Energy 2000 Committee's citizen volunteers intend to make sure it happens.

CONTACT: *Bill Stirling, Energy 2000 Committee, Stirling Homes, Inc., 600 E. Main Street, Aspen, Colorado 81611; 303-925-5757. James Udall, Director, or Lynne Haynes, Assistant Director, CORE, PO Box 9907, Aspen, Colorado 81612; 970-544-9808.*

**DAVIS, CALIFORNIA**　An excellent example of how energy conservation and renewable energy use can be firmly integrated into both city laws and public attitude can be found in Davis, California. This college community, home of the University of California at Davis, operates a host of nationally recognized programs that include solar access laws, minimum insulation requirements, and installation of energy-conserving devices when homes are resold. The city also offers buyers and builders free use of blueprints for solar single-family and duplex

*Creating a Community Value System*

*Strong community support, fostered over twenty-five years, puts Davis, California, on the cutting edge of energy efficiency and renewable energy programs. These bicyclists, moving on the green light from the first-in-the-country bike traffic signals, are travelling at the most congested intersection in the city. At peak times, over 2,000 cars, 1,100 bicycles, and untold numbers of pedestrians move through this intersection each hour.*

Photo credit: Timothy Bustos/City of Davis.

homes, requires that commercial and public parking lots be landscaped so that 50 percent of the paved area is shaded, in order to reduce the heating effect of the parking lot on adjacent areas, has converted its automobile fleet to high-mileage cars; and strongly encourages bicycle use by building bicycle lanes, improving bicycle access to buildings, and installing the country's first bicycle traffic signals.

There's no question that the climate fostered by UC Davis students and faculty, who make up one-third of the city's population, gave Davis its early start on energy efficiency and renewable energy. Such "liberal" programs, however, would never have survived and flourished without strong community support—from residents, city officials, planners, builders, and even the local newspapers. This emphasis on energy, begun in the early 1970s, continues today as Davis pushes ahead with an ambitious bikeway plan. This long-lasting effort has gradually moved energy from a special issue to a basic part of town business and attitude:

> Peer pressure within the city has also helped provide incentives to live energy-conserving lifestyles. More and more people are riding bicycles and driving small cars. The advent of a Cadillac or Lincoln or other large luxury car in a Davis neighborhood would probably be the cause for serio-comic teasing. The creation of a community value system which the majority supports can easily become stronger through unexpected internal actions and judgments such as these.*

CONTACT: *Cynthia Basinger, Administrative Analyst and founding member of City of Davis GREEN, or Timothy Bustos, Bicycle/ Pedestrian Program Assistant, City of Davis, Public Works Department, 23 Russell Boulevard, Davis, California 95616; 916-757-5686.*

## Cree District Heating

**OTTAWA, ONTARIO**    After more than a century of being shuffled from place to place and living in temporary villages, the Oujé-Bougoumou Cree people finally negotiated with the Canadian government for a permanent village of their own. Tribal leaders wanted to build an efficient, district heating system powered by waste wood from local mills. But federal officials refused to fund even a feasibility

---

*City Energy Conservation Programs, Policies and Ordinances, City of Davis, California, 1984, 23.

study since the new village was located close to an existing hydroelectric grid which could provide plenty of cheap power for electric heaters.

Members of the tribe wanted nothing to do with electricity from the massive James Bay hydroelectric project, which was responsible for flooding vast areas of their ancient homeland. They also wanted an energy source the tribe could control, to reduce or eliminate dependence on commercial or governmental power. After a long search, tribal elders discovered that the Department of Energy, Mines and Resources (now known as Natural Resources Canada) was investigating district heat systems. The department's Energy Resources Laboratory ultimately funded both preliminary studies and construction of a state-of-the-art district heating system.

Its hub is an automated wood-fired boiler. Plastic pipes radiate outward to the community, carrying hot water to 122 homes, a school, a church, the town hall, a medical clinic, and several commercial buildings. Inside each structure, hot water circulates through radiators equipped with thermostats for temperature control. The system also delivers hot water for washing and bathing, eliminating the need for individual water heaters. The system is attracting attention from both citizens and government officials. It may help change Canada's focus from massive central power plants to smaller, community-based plants.

If the Oujé-Bougoumou band had based its energy decision on immediate economics only, electric baseboard heaters would have been the least expensive option, said Chief Abel Bosum at dedication ceremonies for the plant. But taking a long-term view, the district heating system offered overall reduced energy consumption, an enormous potential to control cost increases, energy dollars staying in the community and contributing to economic development, and the potential for

*Aerial view of the new Oujé-Bougoumou Cree village in northern Ontario, where all of the buildings are connected to the district heating system. At the hub is an automated wood-fired boiler; plastic pipes radiate outward to homes, businesses, and government buildings, carrying hot water for space heating and washing.*

Photo credit: Guy Paré/Snow Goose Productions.

*Oujé-Bougoumou's decision to install a district heating system combines traditional Cree values with contemporary technology. In their words, they are "generating energy as if people mattered."*

Photo credit: Guy Paré/Snow Goose Productions.

heating energy self-sufficiency. "Once we had satisfied ourselves that the initial capital costs could be repaid relatively quickly, we felt we had no choice but to go in favor of the long-term community benefits by proceeding with the district heating system," he said.

CONTACT: *Paul Wertman, In-House Advisor for Oujé-Bougoumou Crees, Grand Council of the Crees, 24 Bayswater Avenue, Ottawa, Ontario K1Y 2E4, Canada; 613-761-1655.*

## Energy-Efficient House Plans

**CALIFORNIA** Passive solar and energy-efficient homes built without clear, well-tested plans often run into trouble. They can be either too hot or too cold in the winter, and positively hellish in the summer. Yet finding examples of well-tested plans for affordable housing can be difficult. While little initial design money may be available, hiring experienced architects or designers can add thousands of dollars to the cost of a project. Not only would this money be better spent in construction, but cost often prevents solar and energy-efficient features from appearing in affordable housing.

To overcome this barrier, the California Energy Extension Service contracted with the Habitat Center to produce an inexpensive catalog of energy-efficient house plans. Although the plans were designed to meet specifications for housing on Indian reservations, they are suited to the entire affordable housing market.

The ten different design packages include builder-ready plans customized for California climate zones. The $150 package can save up to $1,800 or more on home design, money that can be put into construction instead. The plans incorporate passive solar technologies such as proper orientation, lots of south-facing windows, clerestories, shade overhangs, and thermal mass floors. They also use plenty of insulation to keep heat inside, and other energy efficiency measures such as double-glazed windows, energy-efficient lighting, and high-efficiency wood stoves and appliances.

So far, 150 homes have been built throughout California using the pre-packaged plans, with equal success in the high desert, valley, and coastal regions. Twenty-one homes are located in the Benton Paiute community on the high desert between the Sierra Nevada mountains and the Nevada border. Their owners burn two to three cords of wood a year, and bills for electric backup heat average $35 a month during the winter. In the Klamath River valley in northwestern California, seventeen homes have been built using Energy Extension Office plans, and twenty-four more are planned. Because the climate there is more overcast, and not all the homes could take advantage of southern orientation, the plans emphasize wood stoves for heating instead of passive solar design. Additional successful efforts include: Manzanita Reservation, Hopland Rancheria, Yuraok Reservation, Smith River Rancheria, Picayune Rancheria, Elk Valley Rancheria, and Colusa Rancheria; other tribes have plans on order.

Audrey Flower, executive director of the Karuk Tribe Housing Authority in the Klamath River valley, says the plans fill an important need. "When I saw mention of plans available at such a low cost, I thought it was too good to be true, but worth a look. When I reviewed the plans, I was excited not only by their energy efficiency, but was truly impressed with the open, liveable floorplans and overall quality of design."

CONTACT: *Bonnie Cornwall, California Energy Extension Service, 1400 Tenth Street, Sacramento, California 95814; 916-323-4388. Lynn Nelson, The Habitat Center, 162 Christen Drive, Pleasant Hill, California 94523; 510-825-8434. Audrey Flower, Executive Director, Karuk Tribe Housing Authority, 1320 Yellowhammer Road, Yreka, California 96097; 916-842-1644.*

**TAOS, NEW MEXICO**   Rules are sometimes easier to change than people's long-held ideas and attitudes. When it comes to building a home, for example, most of us think of bricks, wood, concrete, and glass. Not Michael Reynolds. He thinks in terms of earth-packed tires— "rubber-encased adobe bricks"—and empty soft-drink or beer cans embedded in a cement matrix. Earthships, as Reynolds calls his home concept, are either built into a hillside or bermed with earth on three sides.

*Earthships*

*Architect Michael Reynolds calls these structures "Earthships." Like a ship, the buildings are meant to be self-contained, independent, and carry the inhabitant to a better future. The homes heat and cool themselves, make their own electricity, catch their own water, consume their own sewage, and grow their own food. Owner-builder Sam Bascom works on this Taos, New Mexico earthship-in-progress.*
Photo credit: Mel Christensen.

The mostly glass, south-facing wall gathers sun for light and heat, while the thick, earth-packed walls store heat and insulate the building. Two prototype communities for Earthship dwellers are growing slowly outside of Taos, New Mexico. Individuals across the country are also building these energy-efficient dwellings, including actor Dennis Weaver. So far, more than one hundred are either inhabited or under construction.

Jean Mills and Carol Eichelberger are completing their Earthship on their organic farm in Coker, Alabama. They expect to use more than one thousand tires for the 2,000-square-foot home, and to spend less than $50,000 building it. "Most people can learn this; we have a lot of people who want to volunteer on it because they are so intrigued," says Mills. "Earthships have great potential in terms of low-income and community housing." Locations of other Earthships include Ocala, Florida; Hesperia, California; Winamac, Indiana; and Carlton, Washington. Solar Survival Architecture in Taos holds seminars on how to build these homes, and offers a variety of do-it-yourself books and videotapes.

CONTACT: *Solar Survival Architecture, PO Box 1041, Taos, New Mexico 87571; 505-758-9870.*

**TUCSON, ARIZONA**    The Tucson Solar Village isn't just changing the rules—it's changing the entire game. This 820-acre development aims to become "a community in the spirit of the Civano period, a golden era in which the Hohokam [an ancient Native American civilization of the desert Southwest] created a culture based on a balance between natural resources and human needs." That means wisely using energy and water, and generating minimal waste.

Tucson Solar Village will ultimately be the home and workplace for five thousand people. Homes will rely heavily on passive solar design and photovoltaics, using significantly less energy than traditional Tucson homes. Wastewater will be reclaimed whenever possible for outside irrigation and toilet flushing, thus cutting water needs substantially. This sustainable community will be relatively compact to encourage residents to walk or bicycle to work, school, or nearby stores, thus reducing air pollution from automobiles.

This ambitious solar community is the brainchild of a wide spectrum of Arizona leaders. The Tucson-Pima County Metropolitan Energy Commission built a close partnership between city, state, and federal agencies, home builders, and citizen groups. Citizens and environmental groups sit on the planning board, helping to shape the village's future and creating broad support for it.

In the current phase of the project, planners have assembled a multidisciplinary team to see if their goals for energy and water use are attainable. They will also identify a variety of options by which builders can achieve these targets. If all goes as planned, the Arizona State Lands Department will put the land up for bid in late 1994 or early 1995. Under the Arizona constitution, the State Lands Department must sell the property at public auction to the highest bidder.

CONTACT: *John Laswick, Project Manager, Office of Economic Development, City of Tucson, PO Box 27210, Tucson, Arizona 85726; 602-791-5093.*

*Civano: A Model Sustainable Community*

**MASSACHUSETTS**    Demonstration projects underway in Massachusetts, California, and elsewhere hint at substantial changes in a sacred American rite: commuting to and from work. Electric vehicles are ideally suited for this purpose. They are quiet, energy-efficient, generally sit for several hours between uses (plenty of time for a full charge-up), and if photovoltaic panels are used to help charge batteries, they

*Alternative Fuels Program*

don't pollute the atmosphere—either from the tailpipe or from the smokestack of a power plant.

To test and promote electric vehicle use and integrate it with public transportation, the Massachusetts Division of Energy Resources is buying twenty electric vehicles from different manufacturers. These vehicles will be leased to commuters who agree to drive to work each day with at least one other person and park at one of three rapid-transit stations. Here, special parking bays will be fitted with photovoltaic-assisted charging devices. (The solar panels will generate electricity and

## PROFILES IN ACTIVISM
### *Mary Tucker*

Our nation's most unsung renewable energy source may be people like Mary Tucker. Her whirlwind schedule, the quiet power of her convictions, and the ceaseless flow of her work could easily replace—and has actually helped cancel—at least one nuclear power plant.

Tucker is currently a planner for San Jose's Environmental Services Department, a driving force behind the city's new solar access guidelines (see page 116), president of the Northern California Solar Energy Association, and chair of the American Solar Energy Society's 1994 annual meeting, to name just a few of her roles. Fifteen years ago she was a single mother with little knowledge of renewable energy, "who wanted a better world for my children." That better world definitely did *not* include two additional nuclear power plants in their home town of Toledo, Ohio. So she joined a campaign to oppose the plants and hasn't stopped promoting renewable energy since. "If you're going to be against something, you also have to give people some good alternatives," she says. For Tucker, solar power makes an excellent energy alternative.

She helped found a Toledo-area solar energy group, worked on a solar action plan for the state of Ohio, and organized for the Ohio Solar Energy Association. After a move to Washington, DC, to direct a national education campaign on high-level nuclear waste, she founded the Maryland-DC Solar Energy Industries Association. "I'm a born organizer; it's in my blood," she explains.

That comment perfectly illustrates Tucker's first principle for exploring a new career: Recognize the skills you have and use them. The second principle is just as direct: Take some risks and join an organization, go to meetings, volunteer to help. Above all, keep on learning so you can take advantage of new opportunities as well as make them happen.

This is clearly the voice of experience. When Tucker began working on solar energy, she knew little about it. And, unlike most of her counterparts, she had never gone to college (something she is now squeezing in between her day job, her work on various state and national panels, and her grassroots organizing on behalf of solar energy). She "became an information junkie," showed up at meetings, and raised her hand when the call for volunteers went out. Not only did her work pay off for the anti-nuclear fight, it also launched Tucker into a career that she

send it into the utility grid even when the batteries are fully charged, as well as on weekends and holidays, when cars aren't being charged at all.) Massachusetts officials estimate that the PV arrays will provide between 50 percent and 80 percent of the electricity needed for the cars' daytime charging. The rest of the electricity will come from the electric grid.

Several states on the east coast recently adopted low-emission vehicle standards first defined by California. Electric vehicles are expected to play an important role in Massachusetts and New York in

loves. Today her impressive resume looks seamless, with one job flowing directly into the next, more responsible position. "But it doesn't include the months of biting my nails and wondering if I was going to get another job," she laughs.

A long-time colleague, Union of Concerned Scientists' Donald Aitken, describes Mary Tucker like this: "Perpetual motion is a physical impossibility, but Mary is the closest thing to it I've ever encountered. I would guess that her most remarkable achievement is that her exhausting pace never leads to compromise in the quality of performance or product."

Tucker is still trying new things and long ago branched out in her professional work. She remains heavily involved in energy issues, and was recently appointed to the Energy Task Force of the Urban Consortium, an organization of the country's forty largest cities. But now she is looking at the bigger picture, trying to help San Jose become a sustainable city. That means dealing with water quality, sewage treatment, air pollution, housing, recycling, and a host of other issues. "I feel like my brain has exploded in the last year," she jokes. It's the kind of explosion, though, that spreads waves of knowledge, experience, and commitment to an ever-growing number of people.

*Mary Tucker uses her unquenchable enthusiasm and innumerable outreach tactics to achieve her dream of a sustainable energy future, including this organizing staple: coordinating booths and information at San Jose's Solar City Fair.*
Photo credit: Ken Toy.

*Changing the transportation rules is a "must" to reduce energy consumption and improve air quality. California is leading the way with stringent zero-emission vehicle (ZEV) standards, and several Northeastern states may soon follow suit. A demonstration project in Massachusetts uses electric cars like this Solectria Sunrise for commuters who drive from outside Boston and park at a rapid-transit station. There, the car is re-charged during the day from PV panels on the station's roof.*

Photo credit: Solectria.

meeting these standards, as state laws require that 2 percent of the new cars and light trucks sold in 1998 must be *zero*-emission vehicles. This Massachusetts demonstration project will provide crucial data about electric vehicles and commuting. It will also expose a handful of drivers, their families and neighbors, their co-workers, and other commuters to this necessary new form of personal transportation.

A second arm of the alternative fuels program is evaluating how modified state fleet vehicles operate on propane, compressed natural gas, methanol, ethanol, and clean diesel fuel.

CONTACT: *Massachusetts Division of Energy Resources, Alternative Fuels and Electric Vehicle Program, 100 Cambridge Street, Room 1500, Boston, Massachusetts 02202; 617-727-4732.*

# CHAPTER 6

# Strategy #5: Educate by Example

Sometimes a renewable energy project springs from an idea so compelling it just can't be ignored. Volunteers and funds appear as if by magic, the work almost organizes itself, and the end product immediately takes shape. In *most* cases, though, an idea needs a champion, one person or a group of people with the vision, talent, and plain old grit to see it through. Superhuman skills and in-depth knowledge usually aren't required. Much more essential for success are organization, good people skills, and a clear plan, fired with a jolt of passion and tempered with patience. Finding the right combination is tough, but not impossible.

The benefits of turning an idea for a renewable energy project into action, and action into a working wind farm or photovoltaic system ripple far beyond reducing fossil fuel use. Simply bringing people together for a common cause creates a sense of community often missing in large and small cities. More important, such projects teach valuable hands-on lessons in energy planning and technology. They also offer participants the chance to practice and master new skills such as carpentry, electronics, fundraising, or lobbying that they might never have tackled otherwise. Finally, even the smallest project improves renewable energy's visibility, a crucial element for promoting its use everywhere.

As described in Appendix A, solar, wind, and water power technologies have come a long way. Once equated with "flower power" and the back-to-the-land movement, or else with NASA and the military, they are now ready for mainstream commercial use. What holds them back? Manufacturers, retailers, architects, and contractors blame a lack of consumer demand. People just don't ask for these technologies, they say. Nor will such requests become common until people see solar or wind systems in person, on television, or in the newspaper. An interesting, creative project can accomplish this. Raising public awareness that renewables are ready today, and not years from now, can increase demand and nudge the energy industry to offer, perhaps even promote, these technologies.

EarthConnection, an energy-efficient education center, library, and office recently opened in Cincinnati, relies on an often overlooked renewable resource—the human spirit—to promote solar power and sustainable design. The center beautifully illustrates how a project and its organizer educate by example and teach vital lessons on energy, organizing, and the future.

## EarthConnection: From One Woman's Dream to Reality

**CINCINNATI, OHIO**   On a ridge overlooking the Ohio River, Sister Paula Gonzalez is building one of her dreams. She and dozens of volunteers built and transformed a drafty, old, four-car garage on the College of Mount St. Joseph campus into a bright, energy-efficient, 3,900-square-foot community center called EarthConnection. Most of the center's heating and cooling come from a combination of active and passive solar strategies. All the materials selected for the renovation are environmentally friendly. The structural timber frame, for example, uses less wood than a conventional stick-built structure, and requires substantially less energy to manufacture. Recycled products appear everywhere, from the cellulose insulation made out of old newspapers to carpet woven from recycled plastic bottles.

"All of us need to live lightly during our time here on earth. I'm hoping that this building can show people how to accomplish that sometimes-difficult task," says Sister Paula, a 60-year-old biologist, college professor, and self-styled futurist. She also belongs to the Sisters of Charity, an order of Catholic nuns. That might come as a surprise to anyone meeting her during a Saturday morning construction session. Dressed in a WomansWork sweat shirt and blue jeans, her gray hair speckled with drywall dust, she challenges nun-ly stereotypes.

The first Earth Day in 1970 inspired Sister Paula to think seriously about energy and environmental issues, and incorporate them into the biology classes she taught at the College of Mount St. Joseph. Later, when anti-nuclear forces were battling a proposed nuclear power plant in Cincinnati, Sister Paula realized that opposition alone couldn't solve our energy problems. What people really needed, she reasoned, were safe, reliable, low-impact energy alternatives. In lectures, workshops, and essays she's been spreading the message of energy efficiency and renewable energy ever since. "People need to know about and hear

EarthConnection's founding director Paula Gonzalez points out energy-efficiency and renewable energy features on this 1/2-inch scale model built by the design team of University of Cincinnati co-op architecture students.
Photo Credit: EarthConnection Collection.

about alternatives in ways that will be accurate, understandable, and motivating," she says. "We have to reach their moral cores as well as their brains." EarthConnection does exactly that.

This was not Sister Paula's first foray into low-impact design and construction. On Saturday mornings from August, 1982 to July, 1985, she and a band of volunteers turned part of an industrial-sized chicken coop on the Sisters of Charity Motherhouse grounds at Mount St. Joseph into a snug, solar-heated home for herself and another nun. The volunteers scavenged building materials from trash piles and construction waste. They raised money for the project through yard sales, and scoured the city for scrap metal and other recyclable goods, which they sold to local dealers.

This was strictly a learn-as-you-go project. Local contractor Jerry Ropp was so taken by Sister Paula's vision that he showed up every Saturday for three years. Under Ropp's direction, a team of eager volunteers transformed one corner of the chicken coop into a superinsulated, 1,500-square-foot home that gets most of its heat from the sun. Construction was completed just in time for Jerry's wedding to Jean Staas, another Saturday morning volunteer.

The aptly named La Casa del Sol caught the attention of *Mother Earth News* magazine. In a May 1986 feature story, the magazine says

*EarthConnection's energy-efficient windows along the south-facing wall direct sunlight into a sunspace. The incoming solar energy warms the exposed concrete slab, which slowly radiates heat throughout the day.*
Photo credit: Gunnar Hubbard.

the house is "just a first step; it shows that ecologically sensitive living is possible and can even be delightful." *Mother Earth News* also calls La Casa del Sol "a truly exemplary performer. At about 2.5 BTUs per degree-day per square foot for auxiliary heat, it uses half the energy that a top-notch new conventional building does and between a quarter and an eighth of the norm." During a cold winter month, La Casa del Sol uses roughly 500 kilowatt-hours of electricity and salvaged wood scraps for the stove. All this from a two-person house built for under $16,000, or less than $10 per square foot.

La Casa del Sol fulfills a personal pledge Sister Paula made back in 1981 to use less fossil fuels for the rest of her life. EarthConnection represents her more global vision. This energetic nun believes we must somehow bring nature back into our industrial economy and urban lifestyle. Cutting back on fossil fuel use and using more renewable energy is one way of accomplishing this goal. Mimicking natural resource and energy cycles is another. EarthConnection embodies these links by:

- reflecting the earth's natural elements and processes;
- relying directly on the sun's energy and conserving energy wherever possible;
- using recycled, natural, or environmentally friendly materials and construction methods.

For Sister Paula, process is every bit as important as product. So another important goal for EarthConnection is providing hands-on learning for anyone interested in the project, no matter what their level of involvement or experience. In this way, it served as a center for learning long before it officially opened.

From the outside, EarthConnection is a dramatic yet natural structure, blending into its forested site as if it grew there. The bank of south-facing windows draws a visitor in, while the steep ventilation tower, soon to be adorned with photovoltaic panels pulls the eye up toward the heavens. Natural light streams through the wall of windows, making electric lights unnecessary even on gray days. The wide windows look out onto mixed hardwoods and down to the Ohio River.

The rough timber frame further connects the building to its semi-forested environment.

Passive and active solar systems generate the majority of Earth-Connection's heat and all of its hot water. Along the south-facing wall, 433 square feet of low-emissivity (low-E), double-pane, argon-filled windows direct sunlight into a 254-square-foot sunspace. Incoming solar energy warms the exposed concrete slab, which slowly radiates heat throughout the day. Warm air from the sunspace flows into the adjoining meeting room and other first-floor areas. Sunlight beaming into south-facing windows on the clerestory and tower heats and illuminates the building's upper level. Fans circulate some of this warm air down to the lower level.

The 336 square feet of active solar hot water collectors mounted on the south-facing roof over the meeting room play two roles. They generate hot water for the restrooms and kitchenette, and help keep the occupants warm. Solar-heated water flows through seven 400-foot loops of plastic tubing embedded in the concrete floor. Heat from this radiant floor warms people's feet and drifts upward into the rooms.

Northern cities aren't in the best location for year-round solar heating. Cincinnati gets most of its direct sunlight during the hot, muggy summer when heaters are switched off. During the generally cloudy winter, sunlight can generate only a small portion of the necessary heat. An EarthConnection experiment will try to answer a captivating energy question: can July sunlight be stored for January heating? During the summer, water warmed by the solar collectors will be pumped through plastic tubes buried below the center, heating the ground. Warm air drawn out of the building and waste heat from a heat pump's cooling cycle will also be stored underground. An umbrella of buried insulation extending 15 feet beyond the building's perimeter will minimize heat loss from ground to atmosphere. During the winter, in theory, air pulled through the earth beneath the center will absorb heat and help warm the building. Similar systems are operating in Montana, where winters are even colder and less sunny than Ohio. And community-scale seasonal storage systems have been built in Sweden, where they successfully provide winter heat for clusters of homes.

An important part of the dream is yet to be realized: eventually, EarthConnection will sport several photovoltaic panels atop its south-facing tower. Some of the direct current they generate will be used to charge batteries and power special DC appliances. Most, however, will be converted to alternating current in an inverter and used to run standard fans, air pumps, and lights.

Sister Paula and the EarthConnection team are carefully monitoring the performance of these technologies, separately and together. Twenty-nine thermocouples are buried beneath the insulation apron to monitor the heat storage system. Data from these, as well as from several indoor and outdoor air temperature sensors, are linked in a computer. All of this information helps fine-tune the individual energy technologies and integrate them so the entire system works at maximum efficiency. Accurate, scientific monitoring can also supply the kind of data needed to convince a utility or manufacturer that a particular technology, or combination of technologies, really works.

Though the seed for this ambitious project was planted years ago, it began germinating in earnest when the first of Sister Paula's "volunteers from heaven" appeared in 1991. Diane Armpriest attended a five-day spiritual retreat in Colorado Springs, Colorado, where she heard Sister Paula describe La Casa del Sol and talk about her plans for a larger, environmentally friendly center. The nun's dream of integrating natural systems with architecture meshed nicely with the landscape architect's own ideas. So nicely, in fact, that Armpriest took a half-time sabbatical leave for one year from her position as an associate professor of architecture at the University of Cincinnati to lead the EarthConnection design team. Her decision was a crucial step in turning a sketchy idea into wood, glass, and concrete.

"EarthConnection was a chance for me to actively work on something I believed in rather than just talking about it," says Armpriest. She had been looking for a way to shift away from landscape architecture and get more involved with the environmental impact of buildings and energy systems. In more ways than one, this project gave Armpriest the lever she needed to make a change.

True to Sister Paula's spirit of teaching by example and learning by doing, Armpriest involved dozens of University of Cincinnati co-op architecture students in the project. Several helped design the center and investigate appropriate energy technologies. Some focused on appropriate materials for the center, while others actually helped build it. Each gained practical, hands-on experience while moving the project forward inexpensively.

With Armpriest and her team on board, Sister Paula's idea for EarthConnection really blossomed. At a national gathering of the Sisters of Charity in 1991, the order endorsed "healing our global home" as a top priority in its vision statement for the twenty-first century. Sister Paula asked the order's governing council for funds to renovate the rest of her chicken coop. Building a learning center out of mostly recycled materials that used little energy and teaching others to do the same, she suggested, represented appropriate "healing" activities. It also echoed the philosophy of the order's founder, Saint Elizabeth Ann Seton: "Live simply, that others may simply live." The governing board endorsed the general idea. It gave Sister Paula $90,000 and asked that she work on an unoccupied house and garage just up the street from La Casa del Sol on the college campus.

That was in the spring of 1991. By the following fall, Diane Armpriest and her crew had finished EarthConnection's design and working drawings. The city issued a building permit in February of 1992. Demolition and construction began almost immediately after that, mostly on Saturday mornings. The crew consisted of volunteers, University of Cincinnati co-op students paid a nominal wage, and a few paid professionals, many of whom worked for less than their customary rates.

*EarthConnection's design incorporates both passive and active solar systems to generate most of the building's heat and all of its hot water. The solar collectors mounted on the south-facing roof generate hot water and space heating. Solar-heated water flows through plastic tubes set in the concrete floor; heat radiates from the floor to keep occupants comfortable.*
Photo credit: EarthConnection Collection.

These Saturday sessions have a relaxed, homey feel. The regulars pick up where they left off the week before. Newcomers can always find an interesting job, help with new skills or words of encouragement. And everyone gets lunch.

Take a cold Saturday morning in mid-January of 1993. By 9 A.M. the assembled team included Sister Paula, architecture student Jim Herbst, teacher Susan Jones, college student Andrew Baker, La Casa del Sol veteran volunteer Susie Bookser, EarthConnection former executive director Ginny Manss, Union of Concerned Scientists staff member Gunnar Hubbard, and construction managers Ron and Cindy Tisue, who own Solar Bank Energy Systems in Cincinnati. Three projects were scheduled for the day. With power sanders donated for the weekend, one team sanded rough spots on the building's exterior; another ground up chunks of left-over foam panels for use as insulation; a third framed the walls for the center's bathroom. The atmosphere was relaxed, businesslike, and school-like all at the same time. Inspired by Sister Paula's vision, everyone shares a single purpose and cooperates in bringing EarthConnection to life.

This collaborative process "certainly isn't the quickest route to completing EarthConnection," laughs Sister Paula. But what it lacks in speed it more than makes up for in educational value. She regularly gives tours of the work-in-progress to curious passersby, renewable energy aficionados, and even utility company officials. (After one such tour for utility officials, she was invited to join the Demand-Side Management Collaborative of Cincinnati Gas and Electric. The company funded EarthConnection's active solar system, including sixteen collectors, one mile of plastic tubing, and sensors, with a $24,600 grant, and also contributed $7,500 to install state-of-the-art light fixtures, lamps and wiring.) For volunteers, EarthConnection acts as a huge, open-air classroom. Susan Jones, for example, hadn't done much in the way of construction outside of holding tools for her father during home fix-it projects. Now she feels comfortable running a circular saw, has framed doors and windows, and has operated a backhoe. "Working on this project has given me a lot of confidence," she says. Not just in handy skills, but in looking for a new career. She recently left teaching for a job with the Hamilton County solid waste program explaining recycling and waste disposal to teachers and their students, and credits EarthConnection for helping her discover a new dream. Twenty University of Cincinnati students learned the art of timber

*UCS intern Gunnar Hubbard joins other volunteers at the Saturday work session. EarthConnection's construction process is as important as the final product for helping people learn the lessons of conservation, recycling, and renewable energy use.*

Photo credit: Gunnar Hubbard.

framing from a team of masters one weekend. Their final exam was raising the center's interior frame.

The $90,000 grant from the Sisters of Charity, while generous, covers only one-third of Earth-Connection's cost. That means a never-ending round of fundraising and creative financing for Sister Paula and her volunteers. One strategy has been convincing suppliers to slash the cost of building materials, like the expensive double-glazed, gas-filled windows, for which the manufacturer and dealer combined to knock 70 percent off the retail cost. Many needed very little convincing. Cin-Fab, a ventilation contractor, completely re-designed EarthConnection's ductwork and installed top-of-the-line ducts for free, saving the project $7,000. Union electrician Dave Smith was hired to do some wiring. After he was laid off from his job, he returned to EarthConnection and eventually wired most of the building for free. Dayton Water Systems designed, built, and installed a cistern for gathering rainwater after company president Jim Baker heard Sister Paula give a lecture on solar power.

*It's a precise fit! The tower settles perfectly into its base—with no nails in the entire structure. Sunlight streaming into these soon-to-be south-facing windows will provide heat and light dispersed throughout the building.*
Photo credit: Randy Sizemore, Entropy, Ltd.

Used or cast-off goods also help keep cash costs down. For nothing more than a two-hour truck trip and the cost of repairs, Earth-Connection got forty-nine solar water heaters the University of Dayton had obtained when it bought an old building. Sister Paula raises money for the project as she tools around Cincinnati and its suburbs in her red pickup truck, forever on the lookout for scrap metal, aluminum, and other salvageable discards. She sells some of the goods to local scrap dealers, while volunteers sell other items in an annual giant yard sale.

With a little luck and a steady flow of funds, EarthConnection opened its doors at the end of the year 1994. Energy experts and futurists—Sister Paula included—teach classes and hold workshops in the meeting room. Interns, students, and visitors find the library packed with the kind of information needed to design their own renewable energy projects or educate themselves about other environmental issues. A small store showcases and sells an array of recycled goods. In short, the building is abuzz with activity, all springing from one woman's idea.

*Paula Gonzalez discussing exterior sheathing on the building's north side with one of the construction managers. Ron and Cindy Tisue oversaw the project to completion, sequencing each construction step and supervising the host of volunteers who showed up every Saturday to lend a hand. EarthConnection was educating by example long before the building opened its doors.*
Photo credit: Gunnar Hubbard.

When Sister Paula looks at the world, she sees the Word of God proclaimed in water, rocks, trees, birds, and people. In much the same way, when visitors look at EarthConnection, they'll be seeing the ideas of Sister Paula proclaimed in timber frame, solar design, and recycled materials.

CONTACT: *Paula Gonzalez, Founding Director, and Donna Bessken, Executive Director, EarthConnection, 370 Neeb Road, Cincinnati, Ohio 45233-5101; 513-451-3932.*

**LESSONS LEARNED** EarthConnection began as one woman's dream—a building that would teach others about living lightly on the planet, all the way from design and construction through years of use. Sister Paula Gonzalez managed to convey her idea clearly to supporters and volunteers, making it easy for them to develop the reality. For a project's ultimate success, such clarity coupled with commitment, organization, and cooperation may be far more important than a charismatic founder.

The EarthConnection team never hired a fancy public relations firm to promote its project. Yet the message spread far and wide. Sister Paula mentioned it whenever she gave a seminar or lecture. Simple news releases caught journalists' attention. Even the annual giant yard sales promoted the new building and the ideas it embodied.

Furthermore, work on EarthConnection and La Casa del Sol wasn't limited to experts. Anyone with the interest, curiosity, and desire to get out of bed on a Saturday morning was welcome. An open-door policy like this makes it possible for people who know nothing about construction or renewable energy to learn while doing. There's little gained by preaching to the choir—spreading the message of renewable energy means reaching out to people who know very little about it. Sometimes new converts make the best preachers.

Manufacturers continually cite lack of consumer demand as one reason for failing to offer energy-efficient products or renewable energy technologies. Yet, most people don't even know they exist, so they can't possibly ask for them. A project like EarthConnection simply and quietly puts both energy efficiency and renewables technology into a real-life context, so everyone can see how they work and learn about their benefits.

Renewable energy projects are often limited only by an individual's or group's imagination and money. For maximum educational impact, high visibility always helps, like photovoltaic panels atop a fire station that everyone passes, or solar water heaters on a school building. At the same time, a scheme should be matched to a community's needs and should not try to force renewable energy someplace where it clearly isn't welcome. The projects described below give some sense of the range of educational and renewable opportunities that have already proven successful.

**ALAMO, TEXAS**    Down on the flat, fertile plains of south Texas, hard by the Rio Grande, farm workers often live in ramshackle homes made from bits of castoff plywood, sheet metal, and lumber. They're little more than dark boxes that keep out the rain—sometimes—rather than what most people would consider a home. Some have no electricity, running water, or indoor toilets. Clusters of these houses are called *colonias.* Colonias are rural subdivisions, often with no paved streets, municipal sewage, schools, police, or fire services. Throughout the Southwest and Mexico, colonias provide shelter for tens of thousands of people.

Proyecto Fe y Esperanza (Project Faith and Hope) aims to help farm workers build something better. Members of the Proyecto have designed a low-impact home that relies on simple technology for cooling, heating, cooking, and waste disposal. The plans call for orienting homes to catch the winter sun for heating and avoid summer sun for cooling; for installing composting toilets and simple solar water heaters; and for planting vine-covered walls for shade. Just as important, the Proyecto helps workers buy land for their homes, rather than living under the constant threat of eviction.

This project, like so many others, began with an energetic, committed individual who knew little or nothing about home building or sustainable technology. Sister Rosemary Cicchitti arrived in the Rio Grande Valley in 1986 as part of a new mission established there by Catholic nuns of the Belgium-based Missionary Sisters of the Immaculate Heart of Mary. They founded Proyecto Fe y Esperanza (PFE) to improve the lives of farm workers through education and community development.

One of the missionaries, Sister Therese De-Coninck, organized a workshop on technologies appropriate for the colonias and refugee centers across the Southwest. Workshop participants

*Proyecto Fe y Esperanza*

*Utilizing the harsh south Texas climate, appropriate technology advocate David Omick helps colonias residents build solar ovens; this model has a temperature gauge inside. As with all their technology ideas, this solar oven was successfully tested first by Project staff.*
Photo credit: Kevin Gallagher.

included Sister Paula Gonzalez, founder of EarthConnection in Cincinnati, Ohio; Pliny Fiske, founder of the Center for Maximum Potential Building Systems in Austin; and the Reverend Albert Fritsch, from Appalachia-Science in the Public Interest.

Sister Rosemary immediately realized how people living in colonias could benefit from the ideas spun out in the workshop. Aided by a grant from her order, she traveled across the United States, learning about sustainable lifestyles and building technologies. She visited Appalachia-Science in the Public Interest in Livingston, Kentucky; Genesis Farm in Blairstown, New Jersey, a center for sustainable agriculture; a Benedictine Monastery in Kansas known for its work linking ecology and spirituality; and the Rocky Mountain Institute in Snowmass, Colorado, an internationally recognized center for energy efficiency.

On her return to Texas, Sister Rosemary and the other sisters bought five acres of land—bordering a colonia—with a large two-story house outside of Alamo, Texas. This small farming town sits a few miles north of the Rio Grande, and about 60 miles west of the Gulf of Mexico. The nuns hope that their sustainable lifestyle work in the nearby colonia may become a model for colonias elsewhere in Texas, California, and the Southwest.

In colonias, farm workers pay high prices for small lots. The traditional arrangement, called a contract for deed, gives the buyer no equity at all. Typically, a developer buys several acres of land and subdivides them into lots. These are sold to farm workers, who pay up to $200 per month over ten years and don't receive title to their land until the last payment is made. Missing a few payments, even during the final year, can mean eviction, with nothing to show for years of hard work. Thus, the colonias are like a leaky bucket—money comes in, but leaks out almost as quickly. PFE is "patching" the bucket through a revolving loan program funded by the Hilton Foundation and private donations. PFE began the colonia home site purchase program to extend low-interest loans to colonia residents to pay off their lots. Colonia residents gain title to their land and have immediate equity.

Appropriate technology homes are another way to "patch the bucket," but must be proven to work in the harsh south Texas climate. That's where David Omick comes in. Omick, once a sail maker in Brownsville, Texas, left his job to study new technologies for developing more sustainable construction methods, and other ways to live less expensively and more lightly on the land. He built his own low-impact home in Brownsville, which Sister Rosemary learned about through the

Project Faith and Hope's model colonia home was designed to be comfortable, affordable, and appropriate for climatic conditions. In the spring and summer, the strings will support loofah vines used for shade and sale. In the background, the first farm worker to build such a home prepares straw bales for insulation. Hopefully, other farm worker families will soon follow his lead.

Photo credit: Kevin Gallagher.

alternative energy grapevine. The nuns invited Omick to coordinate the appropriate technology program in Alamo.

He built a model colonia home and workshop near Casa Juliana, the large house named for a deceased sister that serves as the Proyecto's center of activity. These two buildings start with an innovative, low-cost cement floor, on which stud walls sit, topped with a sheet-metal or shingled roof. The stud walls are wrapped inside and out with chicken wire, stuffed with straw for insulation, then stuccoed. This construction technique makes for easy expansion as a family grows. At $5 to $6 per square foot (less if recycled or discarded materials are used in construction) these simple, yet attractive, colonia homes cost less than standard balloon-frame construction.

The white walls and roof reflect sunlight rather than absorb it. During the spring and summer, climbing vines cover the south-facing walls. Not only do they keep sunlight off the wall, but they grow rough gourds (loofahs) that can be sold for bath sponges. Graywater from the house is used to water these vines. A row of native trees planted by the National Audubon Society protects the house from hot summer winds. A solar oven is often used for cooking meals, and simple solar water heaters provide hot water for washing and bathing. A composting toilet converts human waste into fertilizer for fruit trees, and composted kitchen scraps fertilize the vegetable garden.

This one demonstration colonia home may attract some attention from local farm workers, but it certainly won't spread the concept far and wide. So PFE, in a joint project with Texas Rural Legal Aid and Proyecto Azteca, built two other sustainable homes for the United

Farm Workers, an organization with a strong presence in the Rio Grande valley and wherever farm workers collect. They will house UFW volunteers and, if successful, may become part of the union's own home-loan program.

The *real* test, however, may rest with Isaias Carranza, a farm worker living in the colonia near PFE. He helped Omick build the model appropriate-technology home; and now Carranza, with his wife Andrea and two of their children, will be the first farm-worker family to build their own sustainable home in the colonia. Once the Carranza family's neighbors see the advantages of living in a comfortable, low-impact, renewable-energy–using home, so the thinking goes, they will be inclined to follow his lead.

CONTACT: *David Omick, Proyecto Fe y Esperanza, RR#2, Box 133A, South Tower Road, Alamo, Texas 78516; 210-702-0524.*

## Solar Eternal Light

**LOWELL, MASSACHUSETTS**   The eternal light shining through a Star of David sculpture above Temple Emanuel's ark uses no utility power, no oil, or candles. But it's no miracle, either—two 9-by-13 inch solar panels on the temple's roof feed electricity to two 6-volt industrial storage batteries, which power the light. During the spring, summer, and fall, they even provide enough energy to illuminate the Torah-containing ark itself.

It's a perfect match between energy source and application, says Rabbi Everett Gendler: "The eternal light symbolizes the dependability of the divine and the sustainability of the source of life and energy.

*Rabbi Everett Gendler on the roof of Temple Emanuel in Lowell, Massachusetts, with the silicon solar panels that have been generating power for the congregation's Eternal Light since 1978.*

Photo credit: Arthur Pollock/The Lowell Sun.

This light shouldn't come from fossil fuels that disappear, nor from nuclear power that is life-threatening." He and a crew from his Lowell, Massachusetts, temple rigged up the system in time for the winter solstice in 1978. Ever since then, the congregation has held a special ceremony each December to celebrate the light. They even held a bar/bat mitzvah on its thirteenth "birthday."

The light is just one of several energy and environmental projects at the one-hundred-family temple. Rabbi Gendler marks the solstices and equinoxes with special services, and children plant and harvest rye and wheat each year. Members of the congregation also recently built a cold-frame greenhouse against the temple's south-facing side. They plan to grow greens throughout each winter to decorate the temple.

CONTACT: *Rabbi Everett Gendler, Temple Emanuel, 101 W. Forest Street, Lowell, Massachusetts 01851; 508-454-1372.*

**SHELBURNE, ONTARIO**   Following the best natural design principles, the Toronto Board of Education made sure that form follows function in its new Boyne River Ecology Center. The rural center, which teaches students from Toronto about ecology and natural science, gets most of its heat from the sun through its passive solar design, and generates all of its electricity with a small wind turbine, a small hydropower plant, and several photovoltaic panels.

The center's circular shape is the most efficient design for taking advantage of central heating. Its sod roof insulates the building naturally in all seasons. The north side is built into the side of a hill, so cold winter winds don't pound into it. The south side contains 65 percent of the building's windows, allowing sunlight to pour in during the winter and provide most of the heat and light. The school's wastewater is purified by an innovative, solar-powered "Living Machine," designed by John Todd of New Alchemy Institute fame. Bacteria, protozoans, algae, duckweed, and higher plants break down sewage and recycle its nutrients, and produce clean water.

Students will do more than just watch all these systems at work. They live at the center for a week at a time and become intimately familiar with how well the design keeps them warm and takes care of their wastes. Students spend half of each day inside, studying the renewable energy and conservation concepts embodied in the building. The solar-powered wastewater treatment plant, for example, offers students an opportunity to explore water pollution, water quality, and the interaction of plant and animal communities in recycling energy and

*Boyne River Ecology Center*

matter. They spend the other half of each day in the woods or around the area's ponds and streams.

CONTACT: *John Fallis, Vice-Principal, Boyne River Natural Science School, Rural Route 4, Shelburne, Ontario LoN 1S8 Canada; 416-857-4160.*

## South Dakota Discovery Center and Aquarium

*Adriane Wegman helps her dad design exhibits at the Discovery Center.*
Photo credit: Steven Wegman.

**PIERRE, SOUTH DAKOTA** Children and adults wandering through the South Dakota Discovery Center and Aquarium in Pierre can watch and study the center's innovative heating and cooling source—drinking water flowing through city water mains. Five water-to-air heat exchangers take heat from this water during the winter and use it to warm the 9,000-square-foot center. (The exchangers can pump out a combined 1.5 million BTUs an hour.) The water enters the heat exchangers at 60 degrees F and is returned to the mains at 58 degrees F. It is eventually used in someone's home or business for drinking, cooking, cleaning, flushing, or whatever. During the summer, the exchangers take heat from the center's air and dump it into the water mains.

The entire system is visible and its parts clearly labeled so the estimated fifteen to twenty thousand people who view it each year can understand how cool water can be used for heating. "That means quite a number of citizens making future energy decisions will be more aware of systems like this," says Steven Wegman, former alternative energy director of the Governor's Office of Energy Policy.

CONTACT: *Steven M. Wegman, South Dakota Public Utilities Commission, State Capitol, Pierre, South Dakota 57501-5070; 605-773-3201.*

## 4H Energy Camps

**SOUTH CAROLINA** "Amp Camps" held each summer for South Carolina's 4H Club members aim to teach children about energy safety. Since the programs are funded by three utility companies, the curriculum mostly focuses on traditional energy sources. Campers make battery testers and lamps and learn how power plants generate and transmit electricity. Because of questions from campers, the curriculum now includes lessons on renewable energy and making solar hotdog cookers. And when old buildings at the camps—one in Aiken and the other in Santee—needed renovations, campers helped add passive solar features that have reduced heating and cooling costs.

CONTACT: *Gerald Christenbury, 224 McAdams Hall, Clemson University, Clemson, South Carolina 29634-0357; 803-656-4085.*

**WEST BRANCH, IOWA** An active solar space heating system built for a school gymnasium became the centerpiece of an innovative energy curriculum, thanks to teacher Conrad Heins, now a professor at Jordan Energy Institute in Comstock Park, Michigan. The 2,500 square feet of flat-plate collectors turn light into heat whenever the sun shines. This heat can be used right away to heat an 8,000-square-foot gymnasium, or to heat water for showers or the school laundry. Or it can be stored for later use by warming beds of crushed rock that store the heat and slowly reradiate it. Because the system generates more heat than could routinely be used in the gym, the excess is shunted to a 6,000-bushel silo for drying corn grown on the school's grounds. During an average winter, this active solar heating system supplies approximately 140 million BTUs of energy that would otherwise have to come from fossil fuels.

Heins used the solar heating system as a teaching tool in a class called "Earth, Air, Fire, and Water." The students in this project-oriented introduction to renewable energy analyzed the operation and efficiency of the system. They also built a solar apple dryer and parabolic reflectors capable of cooking food. They modeled the reflectors in algebra class as they learned about focal points. The students also built a vertical solar collector on the south wall of the school's pottery building.

When Heins left Scattergood Friends School to teach at the Jordan Energy Institute, renewable energy evaporated from the curriculum. The solar collector gradually deteriorated, and was shut off in 1989. Unfortunately, even the most innovative project may falter when its champion moves on. Ideally, she or he should involve as many people as possible in both the vision and the effort, so a curriculum or a solar heater or a community energy program becomes important to many others.

In retrospect, Heins now believes the solar space heating system was too large and complicated. However, despite conceptual and practical problems, the project may have faltered but has not failed. The school began renovating the solar collector in 1993, replacing corroded seals and rebuilding the frame from steel rather than plywood.

## Scattergood Friends School

*For students at the Scattergood Friends School in West Branch, Iowa, hands-on experience with renewable energy was all part of the class work in a course called Earth, Air, Fire and Water. Here a student works on a solar apple dryer.* Photo credit: Conrad Heins.

And the school is now considering appropriate sizing and redesign of the system.

CONTACT: *Dr. Conrad Heins, 215 E. Muskegon Street, Cedar Springs, Michigan 49319; 616-696-0603. Jerry Henderson, Maintenance Director, Scattergood Friends School, 1951 Delta Avenue, West Branch, Iowa 52358; 319-643-7600.*

## Harmony Homestead

**SLIPPERY ROCK, PENNSYLVANIA**     When the Alternative Living Technology and Energy Resource (ALTER) Project at Slippery Rock University needed a combination laboratory, classroom, and housing for graduate students, there was little question about what form it would take or how it would be built. After all, ALTER graduates earn master's degrees in sustainable systems, specializing in the built environment, permaculture design, resource management, and agroecology, which combines traditional approaches for studying agricultural problems with the methods and concepts of ecology. Thus a drafty old farmhouse on seventy-six acres near the edge of the Slippery Rock campus was reborn as Harmony Homestead.

"The house is a test bed for sustainable systems issues. It's largely a social experiment in self-sufficiency and the completion of cycles," says Robert Kobet, who designed and directed the reconstruction as his master's thesis. Students, faculty, and interested volunteers did virtually all the gutting and construction. The building is superinsulated and well ventilated to ensure healthy indoor air quality. A twenty-one-panel photovoltaic array produces electricity during the daytime; electricity not used inside the Homestead is fed into the utility grid. A Copper Cricket™ solar water heater generates most of the hot water, and a composting toilet saves at least 40,000 gallons of water a year. Eventually, graywater from the kitchen and bathrooms will be used to irrigate gardens and planting beds around the house.

Harmony Homestead, which was occupied in the summer of 1990, is home to three students and acts as a living laboratory for forty more. The building is continually being redesigned, taken apart, and put back together to test ideas about sustainable construction. It is a prime attraction for people touring the Slippery Rock campus, and its design sparks visitors' interest. Kobet, now an architect and assistant professor in ALTER's built environment track, says he has already worked with two people who intend to build similar sustainable homes.

Funding for Harmony Homestead came from several sources. The Pennsylvania Energy Office contributed $55,000 because the pro-

*Once a drafty old farmhouse, this building on the Slippery Rock University campus has been transformed into a model sustainable building, including a twenty-one panel photo-voltaic array and a solar water heater. It is a living experiment in sustainability for the students, and a prime attraction for campus visitors.*
Photo credit: Chris Leininger.

ject represented an interesting direction in sustainable housing. The school itself furnished $25,000, and State Senator Tim Schaeffer came up with another $5,000 through a legislative fund. Volunteer labor and expertise kept the cash cost of the conversion to $70,000, though the building is now worth twice that amount.

> CONTACT: *Robert Kobet, Assistant Professor of the Master of Science in Sustainable Systems Program, 006 Eisenberg Classroom Building, Slippery Rock University, Slippery Rock, Pennsylvania 16057; 412-738-2957. (Kobet is also assistant professor of Architectural Engineering Technology at Pennsylvania State University.)*

## *Solar Water Heating*

**BEAVER ISLAND, MICHIGAN**     What began as a simple telephone request about a Lake Michigan island's wind potential is turning into a comprehensive energy plan and some inexpensive solar water heaters for twenty-five homes.

Beaver Island sits in the upper end of Lake Michigan, between the mainland and the Upper Peninsula. About 400 to 450 people live there year-round, and the population on this relaxed vacation spot swells during the summer. Electricity for the island comes from the mainland via precarious underwater cables. Growing demand throughout the 1970s and 1980s meant brownouts and intermittent power failures, especially during the economically important summer tourist season.

Island resident Chuck Hooker called the Michigan Public Service Commission back in 1987 trying to find out if the island could support a small wind farm for generating electricity. He finally reached information specialist Tom Stanton, who said that wind might work, but the best bet was to begin with a long-range energy plan. Stanton, who is an adjunct professor of Resource Development at Michigan State and Western Michigan universities, got his fellow faculty members and students, plus those from Jordan College and Central Michigan

*Jordan Energy Institute students help install solar panels for hot-water heating at Beaver Island's Bluebird Bed and Breakfast. Direct experience with solar technologies proves their reliability to the next generation of builders and architects.*

Photo credit: Susan Barnes.

University, involved. This large team is working with many island residents to develop an energy plan—one that will cost the islanders hardly any money.

Students are performing most of the work on the energy plan. They are collecting data on temperatures, wind speeds, and sunlight levels on Beaver Island throughout the year. They are calculating energy use in homes and other buildings. They are also making an inventory of all renewable energy possibilities on the island. Their preliminary plan points out that "the real and immediate problem for the Island was a serious energy trade imbalance. High-priced energy products [are] imported by the Island, while energy payments and jobs are 'exported' to the mainland." The plan identifies conservation and energy efficiency as the first order of business, with renewable energy development to follow.

Already some Beaver Island residents are weatherizing buildings and wrapping water heaters with a vengeance. Their energy plight appealed to the City of Ann Arbor, which, because of an industry demonstration project, owned fifty-seven solar thermal collectors that were gathering dust in a warehouse. The city sold island residents panels for $40 apiece, a $400 savings over the cost of new ones. Those residents who agreed to work with the project's students got a discount to buy solar water heaters for under $900 installed. Based on annual cost-of-energy-savings of $250 to $300 per year (for a building used year-round), this represents a three- to four-year payback. In comparison, the payback period for a system built with brand-new panels would be closer to ten years. "Try to find another reasonably safe investment with a 10 percent return," says Dave Lankheet, a solar contractor who has been teaching Beaver Island residents to install solar water heaters.

Not only do these devices save residents money, but they also help ease the chronic electric power shortage. In most cases the water heaters replace electric models and thus will reduce daily demand.

While Beaver Island's foray into energy activism began with one resident's interest, it is largely motivated and carried out by a host of students, faculty, and professionals, working side by side with Island residents. Through their efforts and enthusiasm more residents are becoming inspired to get involved, and it is these committed islanders who will ultimately determine their town's energy future.

CONTACT: *Tom Stanton, Adjunct Professor, Department of Resource Development, Michigan State University, East Lansing, Michigan 48824-1222; 517-355-3421.*

**MISSOURI** Students at vocational-technical schools in several Missouri counties learned valuable lessons about renewable energy by building "Sun-Way Homes" developed by the state's Department of Natural Resources (DNR). These traditional ranch-style homes, which were then sold by the schools, incorporated passive solar design elements such as modified Trombe walls, south-facing windows, and roof overhangs for summer shading. They also used plenty of insulation and relied on traditional building materials that can be found in any community.

The DNR, in conjunction with regional homebuilders associations, also held a series of two-day seminars that introduced more than 250 builders to the Sun-Way concept; many of the participants became certified Sun-Way builders, submitting plans to the Energy Division for review and approval. Unfortunately, this program was abandoned in the late 1980s when the directorship and the agenda of the DNR changed. (Although the program was dropped, information on Sun-Way Homes and other renewable energy techniques is available to anyone interested.) Similar state-run programs continue in New Mexico and Arizona. In Arizona, for example, the state energy office conducts energy-efficiency training for builders, works with community colleges to train auto technicians to install and troubleshoot alternative-fuel vehicles, hosts workshops for school teachers on building solar ovens and teaching solar energy principles, and holds an annual energy management conference for municipal and nonprofit managers.

CONTACT: *Howard Hufford, P.E., Energy Division, Missouri Division of Natural Resources, 1500 Southridge, PO Box 176, City of Jefferson, Missouri 65102; 314-751-4000. Amanda Ormond, Arizona State Energy Office, Manager of Community Energy Programs, 3800 W. Central Avenue, Suite 1200, Phoenix, Arizona 85012; 602-280-1410.*

*Sun-Way Homes*

## Fully Independent Residential Solar Technology

*Trenton, New Jersey high-school students experience solar technology first-hand at FIRST solar home. Lyle Rawlings, homeowner, solar engineer, and educator, helps young people appreciate the sun's power. These kids are all feeling the sun's warmth on their faces; Rawlings next asks them to estimate the number of watts they feel.*

Photo credit: Invention Factory Science Center's Environmental Design Studio/ Trenton Roebling Community Development Corporation.

**HOPEWELL, NEW JERSEY**     Fully independent solar homes needn't be banished to vacation areas far from the electricity grid. They make perfect sense in the city and suburbs as well, according to the New Jersey-based Fully Independent Residential Solar Technology (FIRST), Inc. The nonprofit company built its first independent house in Hopewell, outside of Princeton, New Jersey. It serves as a living demonstration of renewable energy at work for local schoolchildren, builders, and others interested in alternative energy. It is also home to FIRST founder Lyle Rawlings, a chemical engineer turned energy consultant, and his family.

The 2,300-square-foot house relies on a bank of south-facing windows, high-performance thermal mass walls, and black-tiled concrete for catching and storing heat. Thirty-six rooftop photovoltaic panels generate electricity and charge a bank of deep-cycle batteries which power compact fluorescent lights, an electric stove, television, computer, and other common appliances. A composting toilet reduces water and electricity use (water pumps operate less frequently), as well as greatly reducing the loading on the lot's barely functional septic field. A solar collector provides a substantial fraction of the family's hot water.

Rawlings and FIRST paid for the house by winning a design competition on energy efficiency sponsored by the New Jersey Department

of Environmental Protection and Energy. With the $171,000 prize, plus an extra $4,000 from FIRST, they built the house, which cost approximately $26,000 more than one with conventional heat and electricity. According to Rawlings, with this extra cost rolled into a thirty-year mortgage at 7 percent interest, it adds approximately $173 per month to the mortgage payments, roughly the same as the monthly utility savings.

The solar house has been featured in both national and local media reports. FIRST uses it to teach area residents about solar energy and energy efficiency, including tours for schoolchildren. FIRST also offers potential home builders assistance on incorporating solar energy into their new or remodeled homes, and generally gives advice on other solar or renewable projects.

CONTACT: *Lyle Rawlings, FIRST, Inc., 66 Snydertown Road, Hopewell , New Jersey 08525; 609-466-4495.*

**MARYVILLE, MISSOURI**    Northwest Missouri State University is very close to achieving an ambitious goal: fossil fuel independence. Several projects are playing a key role in this effort. Since 1981, the university has replaced oil and natural gas with sawdust, wood chips, and paper pellets for 90 percent of its fuel needs. By constructing a wood-waste–burning energy plant and converting one of its oil/natural gas boilers to burn pellets made from paper or corrugated cardboard destined for the Maryville landfill, the university saves approximately $405,000 a year for heating, and reduces the amount of waste flowing into the landfill by 36 percent. Further, one hundred retrofitted vehicles now run on an alcohol-gasoline blend, and an ambitious recycling program with five local counties is currently underway.

More than ten thousand visitors have toured the university's energy plant, some from as far away as Southeast Asia, South America, and Europe. The school held a dedication ceremony in September of 1994 to recognize the opening of the paper-pelletizing plant and to promote the idea that fossil fuel independence is not a wild dream, but a sound financial reality. Northwest Missouri State University also holds workshops and seminars on biomass and renewable energy for area residents, engineers, city managers, and politicians.

CONTACT: *Dr. Robert Bush, Vice-President/Director of the Center for Applied Research and the Institute for Quality Productivity, Northwest Missouri State University, Maryville, Missouri 64468; 816-562-1113.*

*Northwest Missouri State University Encon Project*

## Solar Irrigation Pump

**CRYSTAL SPRINGS, MISSISSIPPI**    The irrigation system on a section of the Mississippi Agriculture and Forestry Station no longer consumes fossil-fuel–fired electricity. Instead, it uses energy generated by photovoltaic panels to drive the pumps, which deliver up to 36,000 gallons a day from a small pond. It's a good match, since the irrigation is generally done during the daytime, when photovoltaic arrays are producing electricity. At night, the system uses energy stored from the day's collection in a bank of batteries to aerate the pond water by shooting it into the air.

This solar-powered irrigation system is the first of its kind in Mississippi, a state with few photovoltaic installations of any kind. The station holds regular tours for farmers and interested Mississippians, offering them the opportunity to see photovoltaics at work in an application that could easily be replicated on their own farms or yards. The station also hosts an annual field day, which regularly draws more than eight thousand people.

CONTACT: *Charles Brister, Southern Solar, RR#1, Box 31, Union Church, Mississippi 39668; 601-535-7499. Mississippi Agricultural and Forestry Experiment Station, PO Box 231, Crystal Springs, Mississippi 39059; 601-892-3731.*

## Small Farm Energy Project

**NEBRASKA**    Farming, which once relied on muscle power and horsepower, now consumes plenty of purchased oil, gasoline, and electricity. On small farms, energy accounts for a substantial fraction of operating costs. Nebraska's Small Farm Energy Project, begun in 1976 by the state's Center for Rural Affairs, helped twenty-four farm families build energy-saving alternatives for traditional tasks like drying grain, heating farrowing barns, or disposing of manure. All were inexpensive, home-built solutions that used locally available materials and were easy to maintain. Compared to an equivalent group of farmers, the "alternative" families saved an average of $1,138 energy dollars in 1979, the final year of the test.

Earl Fish farms 380 acres in Cedar County. For years he had used propane to dry corn in his 6,000-bushel grain bin. A solar addition to the bin, which cost less than $500, completely eliminated the need for propane. Fish nailed a frame of hardware store lumber to the bin's south-facing side and covered it with galvanized sheet metal painted flat black. Air drawn from the north side of the bin passes between the collector and the bin wall, heats up, and then is blown through the grain. Don Wuebben built a similar dryer that works without a fan for under $200.

Since 1979, an outreach program has demonstrated energy-saving projects developed by Small Farm Energy Project participants to farmers across Nebraska. Several private foundations, rural electric cooperatives, and feed and equipment suppliers, as well as the Farmers Home Administration and USDA Extension Service, have all sponsored this effort. On-site tours and seminars have generated several similar projects. The builder-farmers are some of the best salespeople. In 1980, a Small Farm Energy Project workshop on manure composting at Ralph and Rita Engelken's farm convinced Martin Kleinschmit that composting wasn't just for hobby farmers or urban environmentalists. For small rural farmers, manure composting saves hauling and disposal costs, replaces energy-intensive chemical fertilizer, and efficiently recycles nutrients. Not only does Kleinschmit now compost manure from his own farm, but he writes and teaches about this simple, environmentally friendly process as well.

The Center for Rural Affairs still sells several copies each month of the *Small Farm Energy Primer,* a detailed guide complete with instructions on replicating the different projects.

CONTACT: *Wyatt Fraas, Project Director, Center for Rural Affairs, PO Box 736, Hartington, Nebraska 68739; 402-254-6893.*

*Energy-saving projects like this solar grain dryer in Belden, Nebraska, help small farmers save money and stay in business. Earl Fish built this simple solar technology for under $500, and completely eliminated the use of propane to dry his corn.*

Photo credit: Center for Rural Affairs.

## The Buildings of Environmental Organizations

Organizations promoting energy efficiency and renewable energy know it's important to practice what they preach. Many have undertaken ambitious office construction or renovation projects that embody the clean-energy message. These buildings serve as working examples that energy efficiency makes sound fiscal sense. They also help teach local zoning boards about renewable energy technologies, paving the way for others to adopt them without facing a bureaucratic struggle.

Each year thousands of people visit the National Arbor Day Foundation's new conference center in Nebraska City, Nebraska. The elegant timber truss and fieldstone building shows off the beauty of wood in construction. More importantly, visitors stepping up to an indoor viewing platform can see the wood-fired heating and cooling system in operation. Wood for this highly efficient biomass gasifier is sustainably grown on plantations surrounding the center; ash from the burner fertilizes the trees. Guides and pamphlets point out that the wood burned is replaced by trees planted, so there is no net input of carbon dioxide into the atmosphere—and thus no contribution to global warming.

The Union of Concerned Scientists has moved into a new building in Cambridge, Massachusetts, that cost the same as conventional construction for the area but that relies very little on fossil fuels. De-

## PROFILES IN ACTIVISM:
### *Reverend Albert Fritsch*

Albert Fritsch is back where he started sixty years ago—on a sustainable farm in Kentucky. But the trip in between has taken the environmentalist, energy activist, and consultant to Washington, India, Africa, and all across the United States.

Today, the Jesuit priest directs Appalachia–Science in the Public Interest (ASPI), an organization he started in 1977 when he split away from the Washington, DC-based Center for Science in the Public Interest, which Fritsch helped found several years before. ASPI's mission, Fritsch explains, "is making science and technology responsive to the needs of poor people in the Appalachia region." The center has staked out three main areas of work and research: renewable energy, organic farming, and low-income housing. Each of these can benefit people trying to stretch out a marginal cash income.

Several buildings at ASPI's Livingston, Kentucky, headquarters sport solar technologies, from photovoltaic panels to simple solar water heaters and passive solar design. Helping people take advantage of the sun's energy while at the same time making their homes more energy-efficient allows them to pay less money for oil, wood, or gas, says Fritsch. And that means more money for food, clothing, and education.

ASPI workers have also designed an affordable composting toilet that can be built for $500. In Appalachia, as in many other rural areas, septic systems are expensive to install and tough to maintain. Homeowners often pipe their sewage directly into nearby lakes or rivers. A good composting toilet costs virtually nothing to operate, protects the watershed, and yields a rich fertilizer.

Even more important is ASPI's work on low-income housing, says Father Fritsch. The center is now collaborating with two lumber companies and an Ohio builder to make low-cost mobile homes that are really permanent. "In a very real sense, most mobile homes today aren't affordable for poor people, even though they cost less to buy than permanent homes," Fritsch points out. Mobile homes are not built to last, require wasteful amounts of energy to heat, and lose their value from the moment they're first occupied. Since many rural Americans can't afford anything else, it makes sense to design these homes so they use energy wisely and last a long time.

The organization makes its ideas available to people through technical papers, building plans, and pamphlets. Many people freely adapt these plans to suit their own needs. Fritsch encourages them to keep records so researchers can study the effects of modifications, but hard-pressed farmers, miners, and factory workers rarely have the time for this.

Fritsch's concern for sustainability and simple living grew out of his boyhood on a farm in Mason County, Kentucky. "We grew everything we needed, including fuel for our horse power. It wasn't until I learned about people getting food stamps during World War II that I realized not everyone lived this way," he recalls. Several

signed by a UCS staff team led by Donald Aitken, the building incorporates nontoxic and recycled materials into a structure with high levels of insulation and efficiently glazed, operable windows to provide a healthy indoor climate while greatly reducing the need for additional heating and cooling. Daylight replaces almost all of the light from the

*Reverend Al Fritsch (right) explaining to visiting college students some of the solar features on a model of a low-cost house being built at the Sand Hill Community Land Trust.*

Photo credit: Appalachia–Science in the Public Interest (ASPI).

years later, the newly ordained priest was finishing up postgraduate studies in chemistry at the University of Texas. He wanted to put his knowledge and skills to work helping the poor, and thought of asking for a position at a Jesuit technical school in Monterrey, Mexico.

"Instead, I got involved with Ralph Nader and his public-interest efforts in the early 1970s and haven't stopped doing that since," explains Fritsch. He began working in the area of toxic metals, like mercury and lead in the environment. That pointed him to work involving lead in gasoline, then to gasoline and energy, then to

renewable energy, and finally to sustainable lifestyles. His next major project with ASPI aims to help an estimated quarter-million Kentucky families find a productive, sustainable alternative to growing tobacco, which has long been a crucial cash crop in Appalachia.

His lively voice gets even more animated when talking about yet another ASPI project, an oral history designed to collect stories about appropriate technology from elderly Appalachia residents. "Many old-fashioned methods for doing all sorts of things were excellent, and they are being lost rapidly. We need to get out there and learn them before the people who can tell the stories are gone."

As director of a pioneering organization, an international consultant, and author of several books on sustainable living, Fritsch has come to understand some of the barriers that limit the spread of renewable energy or appropriate technologies. Institutional rules that prohibit creative, inexpensive technology are one obstacle. The worst, however, may be lack of communication. "There is so much information and so many good plans out there," he says, "but we just can't seem to get them to the people who need it." Father Fritsch and Appalachia–Science in the Public Interest have made that barrier just a little smaller.

*Look at this photo carefully. Notice anything unusual? The Union of Concerned Scientists' new building in Cambridge, Massachusetts sports operable windows. Using this very simple, old-fashioned system, windows that open help provide a healthy indoor climate while greatly reducing the need for additional heating and cooling.*

Photo credit: Herb Rich.

high-efficiency lighting system during both sunny and cloudy days, and PV panels on the roof integrated with the local utility grid provide solar electricity to meet a portion of the UCS office needs. "Utility-friendly" buildings such as this can help utilities reduce their peak demand and put off or eliminate the need for new power plants to support the new buildings. This hypothesis will be verified over the first year or two of occupancy by actual measurements of the performance of the UCS energy and lighting systems by the local utility.

The National Audubon Society transformed a turn-of-the-century building into the most energy-efficient office in Manhattan. Recycled or natural materials were used wherever possible in the eight-story building, which is projected to use 60 percent less energy than it did before the renovation. While all the design decisions for this $14 million project made good environmental sense, each one also offered a one- to five-year payback.

Several other major environmental organizations, including the Conservation Law Foundation, the Environmental Defense Fund, and the Natural Resources Defense Council, have also moved into buildings that exemplify cutting-edge practices in energy efficiency and sustainable design. And all of these organizations are using their buildings to educate their members and others about environmentally sound building practices.

CONTACT: *National Arbor Day Foundation, Mary Yager, Director of Program Services, 100 Arbor Avenue, Nebraska City, Nebraska 68410; 402-474-5655. Union of Concerned Scientists, 2 Brattle Square, Cambridge, Massachusetts 02238; 617-547-5552. National Audubon Society, 700 Broadway, New York, New York 10003; 212-979-3000. Conservation Law Foundation, 62 Summer Street, Boston, Massachusetts 02110; 617-350-0990. Environmental Defense Fund, 257 Park Avenue S., New York, New York 10010; 212-505-2100. Natural Resources Defense Council, 40 W. 20th Street, New York, New York 10011; 212-727-2700.*

# CHAPTER SEVEN

# How You Can Make Renewables a Reality in Your Community

**M**any of the projects outlined in this book were relatively easy to implement, while others were much more difficult. In all cases, however, advance planning and good organization were the keys to the success of the project. This section, along with the Resource Guide (page 213), provide you with the information you need to begin making renewable energy a reality in your community.

For the past few years, the Union of Concerned Scientists has been working to bring renewables off the drawing board, onto the power grid, and into people's lives and homes. The UCS Public Outreach Department has been particularly interested in citizen-initiated, local-level renewable energy projects across the country. We decided that, in order to understand the process activists go through in getting their community-based projects off the ground, we would attempt such a project ourselves. This work helped us understand what is involved in setting up a local renewables project. It also taught us some valuable lessons about dealing with utilities, neighborhood groups, energy experts, and government agencies. In the end, our own initiative resulted in the greater use of energy efficiency and renewable energy in a low-to-moderate-income neighborhood in Boston.

In the spring of 1993, UCS established a partnership with the Neighborhood of Affordable Housing (NOAH), a community development corporation in East Boston, one of the city's oldest neighborhoods. This neighborhood, like many in old cities all across the country, is currently experiencing rapid demographic changes, and it has the highest foreclosure rate in the city. With an old and battered housing stock, and Logan International Airport pressing hard upon the neighborhood, the conditions all conspire to spell disaster for East Boston. But NOAH has a different vision: among other community services it provides, NOAH buys up abandoned, foreclosed properties and ren-

*The installation of high-quality window units was one of several energy-efficiency measures used in the renovation of an old house in East Boston, Massachusetts. This collaboration between the community development corporation Neighborhood of Affordable Housing (NOAH) and the Union of Concerned Scientists demonstrates the feasibility of conservation and renewable energy features in gut rehab affordable housing.*
Photo credit: Kathy Pillsbury.

ovates them, bringing stability and new life to this neighborhood. During the next two to three years, NOAH is planning to rehabilitate sixty to one hundred dwelling units, most of them in three-story buildings, with two to six units per building.

UCS and NOAH realized that, in order for renewable energy applications and energy efficiency to become an integral part of East Boston, the community would need to fully embrace these concepts. Area contractors would need to learn new design and construction techniques, and area utilities would need to provide funding to help offset the initial higher costs involved with the project. We therefore decided to rehab a "model" house that would demonstrate the concepts we were introducing. This model house dramatically proves the benefits of renewable energy and energy-efficient renovations—including lower utility bills—to the very people in East Boston who can encourage such measures: the NOAH board, building contractors, public officials, low-income housing funders, and community leaders. The NOAH example can be replicated in similar communities across the United States.

The centerpiece of our joint project was the renovation of 120 Everett Street in East Boston. UCS's contributions included analyzing the cost-effectiveness of various energy-efficiency and passive solar measures; drafting the specifications for the building contractors; providing technical assistance on the actual construction; gaining cooperation and funding from Boston Electric and seeking support from Boston Gas; and preparing an orientation and operating manual on the home's exciting features for the new residents.

NOAH achieved the energy improvements and savings through extensive insulation and air sealing, energy-efficient lights and appliances, high-quality window units, and a mechanical ventilation system. The renovation also completely redesigned the south-facing front of the house to take advantage of passive solar energy, using the building's structure to capture and distribute sunlight, thus reducing the need for fuel to light, heat, and cool the home. The energy savings are so great that the house's boiler was reduced by two sizes, resulting in construction cost savings of several hundred dollars. The utility bills for this two-family home will be reduced by $300 to 500 annually, and the residents will have healthier living environments—well-lit, well-ventilated, comfortable apartments that are warm and free from drafts in winter, and cool in summer.

How we achieved this success—as well as the disappointments we experienced—is described in this chapter. From conception to completion of our effort, we retrace the steps UCS went through to pick a project and follow it to its conclusion. We believe these are major steps that most groups will need to follow to get a renewables project going: creating a core group, choosing a strategic project, setting goals, conducting audience outreach, harnessing local resources, securing funding, gaining publicity, and following up.

## Creating a Core Group

In UCS's case, the core group already existed: UCS staff members. You may already have a committed core group of your own as well, or you may have to organize one. Except for the very simplest projects, your core group should consist of at least four or five individuals who can share the workload and have a sense of "ownership" over the project. A core group need not be made up of energy experts; for example, the Tuesday Evening Lemonade Club (see Chapter 5) in Austin, Texas, included a former English professor turned housing-cooperative manager, a fiction writer, a journalist, a former nuclear power advocate, and an aide to a city councilwoman. Yet this group showed that a small number of concerned and determined citizens, with little or no expertise in energy issues and with a diversity of backgrounds, can lay the groundwork for sweeping change in their community. What began as a core group of several spirited individuals eventually turned into a renewable energy campaign with thousands of participants.

To create a core group in your community, you need to identify other people who share your concerns. Perhaps you are already a member of a social or environmental action group at your church, synagogue, or school. If not, find out whether national or regional environmental organizations have a local chapter, or whether organizing groups like Citizen Action or a Public Interest Research Group (like MassPIRG in Massachusetts) are active in your area. If you are not already familiar with energy/environmental advocates in your community, you could approach members of the local planning or conservation commissions for names of interested individuals. You can even post flyers in areas with lots of foot traffic, seeking folks to join you in an energy action group.

## Choosing a Project

There are two main steps in selecting a renewables project: determining the strategic opportunities that exist for developing renewable resources in your area, and identifying the resources themselves.

**MAXIMIZE STRATEGIC OPPORTUNITIES**   To be most effective, you should be receptive to strategic opportunities for renewable energy development that already exist in your community—opportunities that are just waiting to be exploited. A "strategic opportunity" is an action that builds upon support that is already available, rather than starting from scratch. Strategic opportunities are important because you can often parlay a small action into a much larger impact, since some of the groundwork has already been laid. Experienced campaign planners often spend extra time and effort looking for that kind of opportunity, as it increases their effectiveness with relatively less effort. In other words, it's often easier and more productive to widen a window of opportunity than to build a whole new door. Look for the following opportunities in your community:

- **State or local incentives for renewable energy development.** The state of Minnesota, for example, has property tax incentives for wind energy development. Your town may a have solar access law. Research how renewable energy use may already be institutionalized in your community, and take advantage of it.
- **High costs of fuel or electricity.** Renewable energy may be more attractive in areas where energy costs are high. The case of biomass utilization in Vermont (see Chapter 4) demonstrates this principle. By conducting a cost comparison between electric heat and biomass (wood chips) heat, the high school's finance manager was able

to show that biomass was significantly less expensive than the existing heating system.

- **A municipal utility.** Municipal utilities are much more likely to be open to citizen input, since they are by definition accountable to the town. As shown in Chapter 2, municipal utilities are often in the forefront of renewable energy use because they often value criteria other than short-term profit.

- **Utilities that have strong conservation programs.** Many utilities in the United States have conservation (often called "demand-side management," or DMS) programs. These utilities will probably be responsive to community interest in renewable energy, because they are convinced, apparently, of the environmental and economic soundness of responsible energy use. Waverly Light and Power, described in Chapter 2, is a good example of a utility that started with aggressive conservation programs and is now pursuing several renewable energy options. And WL&P is not unusual; a national pattern is emerging—utilities with strong DSM programs are most likely to be exploring renewable energy options.

- **Plans for extension of feeder lines, or new distribution substations.** As indicated in Chapter 3, photovoltaics can be cost-effective in cases where feeder or distribution lines need to be extended. If your area utility is planning such an extension (to a new subdivision, for example), you could urge them to explore photovoltaics as an alternative.

- **Public facilities scheduled for renovation.** Buildings account for 33 percent of energy use in this country; public buildings, often old and deteriorating, are notorious energy-guzzlers. If a public building in your community is scheduled for renovation, your group could argue for renewable energy to be incorporated into the building's re-design. There are several stories about building retrofits scattered throughout this book.

- **New residential or commercial construction.** New construction is the easiest and most effective place to introduce energy-efficiency and renewable energy features (easier than, say, a retrofit). The town of Soldiers Grove (see Chapter 5) is an excellent example of exploiting this strategy on a large scale; also see the section on Neuffer Construction (Chapter 1) for another example of renewable energy in new construction.

In addition to the strategic opportunities described in the list above, you might also look to particular strengths you already have in

*Buildings account for 33 percent of energy use in this country, and simple, relatively inexpensive passive solar features can dramatically reduce home energy use—even in old and deteriorating urban housing stock. UCS helped re-design the south-facing front of this model rehab to capture more sunlight for heating and lighting. Notice the window overhangs to prevent overheating in the summer.*
Photo credit: Kathy Pillsbury.

your community, or in those nearby. Any of the following will improve your likelihood of success:

- **Expressed community interest.** Perhaps the most important opportunity to exploit is community will and enthusiasm. As the successful projects in this book demonstrate, if a group of committed people get together and rally around a project, then almost anything is possible.

- **Strong community leadership or potential coalition partners.** If your city council is packed with energy activists, or if your municipal utility board of directors cares about environmental quality in your community, then you've got the leadership and community culture to make something exciting happen. A Public Interest Research Group (PIRG) already canvassing in town on energy issues is an obvious ally with lots of people power. Don't overlook the *people* opportunities at your fingertips.

- **Good probability of success.** Choose a project that you deem has a good chance of succeeding. A failed project may be disheartening to you and may discourage future projects. Yet in your quest for setting and reaching attainable goals, don't be afraid to take risks!

A simple technique for evaluating strategic opportunities is to list your possible program ideas across the top of a page and list strategic opportunities vertically. Check off which programs fit with which opportunities, and pick the program that takes advantage of the most opportunities. When UCS was selecting its project, we initially considered three possibilities. One was providing assistance to the leadership of a suburban town that was very interested in environmental and energy issues; another was to become involved in the renovation of a public building; and the third was the NOAH project. After initial research and careful consideration, we eventually chose NOAH because it embodied the greatest number of strategic opportunities: electricity costs are high in NOAH's community, the area utilities offer extensive conservation programs, there was interest from the community and from NOAH staff, and we projected a high probability of success.

## IDENTIFY PROMISING LOCAL RENEWABLE RESOURCES

The next step in choosing a project for your community is to find out what renewable resources are most applicable in your area. One of the most important things to remember is that renewable energy resources are locally specific. Biomass conversion may be a good energy application for public schools in some areas in Vermont, for example, but not for the EarthConnection community center in Cincinnati. Flat-plate photovoltaics may be a smart choice for Austin, Texas, but not for Waverly, Iowa.

There are several ways to obtain renewable energy resource information. One good place to start is your state's energy office, which may have much of the data you need to identify your local renewable resources. In addition, several resource books provide information about national solar and wind resources:

*An Assessment of the Available Windy Land Area and Wind Energy Potential in the Contiguous United States,* Pacific Northwest Laboratory, Richland, Washington, 1991.

*Wind Energy Resource Atlas of the United States,* Pacific Northwest Laboratory, Solar Energy Research Institute (SERI), Golden, Colorado, 1986.

*Isolation Data Manual and Direct Normal Solar Radiation Manual,* Solar Energy Research Institute (SERI), Golden, Colorado, 1990. [Note: the two SERI manuals listed above were combined into one document and reprinted in 1990.]

*Powering the Midwest,* Union of Concerned Scientists, 1993.

The last resource is available from the Union of Concerned Scientists, 2 Brattle Square, Cambridge, Massachusetts 02238-9105; all the others are available from the National Technical Information Service, U.S. Department of Commerce, 5285 Port Royal Road, Springfield, Virginia 22161.

Finally, the renewable energy industries have lists of resources and local experts who may be able to help you. Such groups as the American Wind Energy Association and the Solar Energy Industries Association can provide you with some of the information you need to get started. (See the Resource Guide, page 213, to find out how to get in touch with these groups.)

Once UCS decided to work with NOAH, we explored the possibilities for incorporating renewable energy and energy-efficiency measures into NOAH's low- and moderate-income housing rehab and renovation projections. In addition to incorporating high-quality energy-efficiency features, we looked at using active and passive solar, as well as heat pumps. Our initial research determined that active and passive solar applications were the most viable options available. Active solar could be used in solar hot water heating or photovoltaics systems, while passive solar building design could bring important benefits. Unfortunately, financial constraints ultimately ruled out the use of active solar systems.

## Setting Ambitious and Attainable Goals

After choosing a project, your group should set clear goals that have a good chance of success. It is a good idea to agree on a series of milestones that can be achieved, one at a time. Meetings, panel discussions, hearing preparations, securing sponsors, and so on, could be short-term steps toward your larger goal. Keep in mind, however, that goals and timelines should always be flexible; situations beyond your control may require making adjustments in your schedule and plans.

As we planned the UCS–NOAH collaboration, we outlined four phases for the project: imparting critical information to all key players, securing funding, constructing the model building, and following up. In each phase, we identified a number of milestones to be reached; the completion of each milestone called for both evaluation and celebration. One milestone in the building construction phase, for example, was the completion of the specs and plans for the house. When they were drawn up, the organizers celebrated their progress; all the initial plans were finally taking form. This success breathed new life into the project for both UCS and NOAH staff.

## Addressing Different Audiences/Forming Coalitions

By the time your group has decided on a project and set the goals necessary for it to be carried out, you will probably have formed some idea of the individuals or groups you want to interact with and audiences you might attract. At this time you should draw up a project description and begin presenting your ideas to those who will need to be involved. During the UCS-NOAH collaboration, we learned that many different groups were involved in NOAH's current operations, and that we would need to recruit others. Those already working with NOAH were local lending institutions and foundations, as well as city, state, and federal government agencies. We found that gas and electric utilities, the East Boston community, local contractors, an energy services group, the Department of Public Utilities, and other area banks would all need to become additional players. For example, the utility companies each had DSM programs that might provide financial and technical assistance to the project; a local energy services group contributed

NOAH/UCS collaborative participants celebrate completion of the specs and plans for the model house. From right to left, NOAH Director of Development Paula Herrington, architect Mark Dulak, Boston Edison Energy Crafted Home specialist Tom Rooney, UCS intern Gunnar Hubbard, and energy consultant David Weitz discuss the plans.
Photo credit: Kathy Pillsbury.

advance technical training and then quality control on the construction site; and area banks had to be educated about energy-efficiency benefits before approving mortgages on NOAH's renovated houses.

To attract interest in our project, UCS made a presentation to the NOAH board, some of NOAH's funding sources, and the City of Boston Public Facilities Department, outlining the project and the utility program as well as other opportunities that NOAH could exploit. In addition to outlining the project, we left time for questions. This meeting allowed us to both present the project and hear about some of the concerns surrounding it. We also met and telephoned many other groups during the planning and implementation of the project: Boston Edison and their contractors, the Department of Public Utilities, the Boston Mayor's office, the Massachusetts Attorney General's office, the National Consumer Law Center, other environmental organizations, and, of course, the architect rehabbing the house. Along the way we kept track of those who praised the project, so that even though their support might not be critical in the initial stages, we could ask them to help later on, if needed.

We believed that the gas and electric utilities could and should be our natural allies in this work, since they sponsor energy-efficiency and renewable energy incentive programs. Although NOAH was aware of these programs, the community development corporation was not participating in any of them. Nor, we discovered, were these programs applicable to NOAH's work rehabbing abandoned buildings: all of the utility programs were designed for new construction or to assist current utility users (which didn't apply to NOAH because, first, the work was rehab; second, there was, obviously, no one living in the house being renovated). So we began looking for ways to get the utilities involved in NOAH's activities.

Fortunately, Boston Edison was very interested in working with us, and the company helped in two major ways. First, they provided us with a computer modeling program that was used to estimate how much heating energy could be saved by different efficiency measures in a building. UCS energy analysts used this program to conduct a cost-benefit analysis of three different sets of progressively stronger energy-efficiency measures.

In addition, Boston Edison stretched the criteria for their highly acclaimed Energy Crafted Home (ECH) program (originally intended only for new construction), so that the NOAH rehab project could benefit. The ECH program helps people design and build an energy-efficient house. Through this program, Boston Edison set up and funded training sessions on energy-efficient construction techniques for NOAH and their contractors, reviewed the plans for the model home, and helped establish procedures so the rehabbed home could be made energy-efficient enough to meet the program's stringent standards. The ECH program also paid to have the building inspected and to have tests performed to verify the efficiency of the construction.

There is more than one way to form an effective coalition or to convince necessary allies to join your project. UCS and NOAH used a go-to-them approach, as described above, where we identified and then recruited the constituencies we needed to successfully complete the project. Another example of this assertive outreach approach can be seen in the Austin story, where organizers set up a volunteer speakers bureau to reach out to the Chamber of Commerce, neighborhood associations, and young people. But Austin also used a come-to-us approach; organizers set up a two-day energy fair—featuring exhibit booths, presentations, and entertainment—which was open to the entire city. This energy fair, commonly recognized in Austin as an important turning point for the city's successful energy work, garnered widespread public support for the organizers' energy policies and helped create consumer demand for alternative energy services. Both approaches—go-to-them or come-to-us—can be very effective, and some projects may require that both be used. You need to determine what is necessary given your local conditions.

No matter how diverse your core and volunteer groups are, you may find that you still lack some of the expertise that will be necessary to complete your project. However, there is a very good chance that someone in your community does have that expertise. At UCS, for example, we needed to give NOAH estimates of the specific energy savings that would result from certain renovations. To do this, we needed to access a particular technical software package and to gain in-depth knowledge of the prices of the newest energy-efficient technologies— neither of which we had. But by contacting many of the people who had expressed interest in our project during our first round of calling, we were able to identify someone who knew energy-efficiency markets and how to use the software.

We greatly encouraged NOAH to incorporate passive solar design into the home's renovations. NOAH didn't have any previous experience with solar design, nor did the architect working on the project; both, however, were very interested. It turned out that UCS intern Gunnar Hubbard was able to provide the design, training, and monitoring expertise to turn the entire front, south-facing wall of the home into a solar collector. Not only will this decrease the heating and cooling load on the house, it also gives the home a spacious, light, and airy feeling—the kind of aesthetic all homeowners should be able to enjoy. The expertise that Gunnar brought to the UCS-NOAH project is available in many communities.

Every location is unique, as are the resources that your group may need. With imagination and persistence, you should be able to tap into the wealth of expertise and goodwill in your community, and come up with the help you need.

Your group may require funding for a wide variety of activities. You may need money for things as diverse as printing flyers, advertising a forum about your project, buying an advertisement in the local paper, or purchasing a wind turbine. Again the rule is: To get what you want, you first have to *ask for it*.

Financial support comes in many forms, including non-monetary. During the UCS-NOAH collaboration, for example, we held a workshop to inform East Boston housing contractors about incorporating energy efficiency into housing renovation. We were fortunate that a local energy services company donated its conference room for the workshop.

*As part of the project's education and outreach, energy consultant David Weitz explains energy-efficient construction techniques to East Boston area contractors at an intensive workshop, held on-site. NOAH specified that only those contractors who took the course could bid on the contract.*

Photo credit: Kathy Pillsbury.

Cold hard cash, however, is often an issue. In NOAH's case, the staff had already secured funding for rehabbing the house; what they needed was additional funding to install the energy efficiency and passive solar measures, which added about $4,000 to NOAH's rehab budget.

We turned to the utilities for assistance, believing that utility involvement in such a project would enhance the companies' stature in the community, as well as provide a needed service to our project. With this in mind, UCS drafted two proposals asking the utilities to help defray the cost of the model home. Boston Edison's help was indispensable but limited, because the real energy savings in the house would come not from electricity but from gas. The Boston Gas energy efficiency programs did not apply to the NOAH project, and we could not persuade the company to bend the rules. Fortunately, NOAH was able to finance the energy improvements from other sources, but lack of support from the gas utility was disappointing. UCS is now working to get Boston Gas and other Massachusetts utilities to modify their energy-efficiency programs to deliver services to those most in need of them—neighborhoods served by community development corporations like NOAH. A fall 1994 ruling by the Massachusetts Department of Public Utilities should eventually make it possible for multi-family gut rehabs to qualify for the gas company's energy efficiency programs.

UCS did help NOAH secure funding in another way, however. Because the NOAH-UCS collaboration was unique and innovative, NOAH was able to attract foundation and government funding from sources traditionally closed to them. UCS staffers also wrote letters of support on NOAH's behalf, and provided information for interested potential funders.

## Gaining Publicity

Publicity is essential for introducing and promoting renewable energy in your community. A successful project usually will be one that gains widespread attention, especially among the audiences you have determined to be critical. You also presumably want to spread your renewables message beyond the scope of your core group, affiliated coalitions, and project participants. You will first want to determine carefully *what kind* of publicity you need, and then *how much* you need. Unfocused attempts to get media attention are usually a waste of precious time.

For the NOAH–UCS project, we determined that the most publicity was needed right there in East Boston. Some members of the community expressed skepticism about the energy-efficiency/solar features: Would they work? Are they worth the extra expense? Others were concerned about the house's appearance. Still others didn't know anything about it. So we focused our publicity efforts on the East Boston neighbors. For example, UCS and Boston Edison sponsored an intensive workshop for area contractors to learn about these techniques, and NOAH specified that only those contractors who took the course could bid on the construction contract. As NOAH uses primarily local contractors, this step served to spread information about renewable/solar energy throughout the neighborhood.

In a different vein, UCS created promotional brochures to help sell the model house, and NOAH distributed them through their extensive community network. We knew that only a few people would actually bid to buy the house, but the brochure was very attractive and explained our work to everyone who saw it. NOAH also sponsored open houses to display the property, assuming many neighbors, as well as prospective buyers, would stop by to have a peek inside. And, on still another track, the project got good coverage in East Boston newspapers that reach both the English- and Spanish-speaking communities in the neighborhood.

The feature stories in each chapter of this book all involve some type of public education and publicity work, and can be re-examined for successful prototypes. The main point is this: there are many different types of publicity; you need to figure out what you really need to advance your project and then look to community resources for help.

## Following Up

An essential but often overlooked step in any project is the follow-up. Follow-up is not glamorous and in fact is often boring, but following up a project keeps the pressure on to change the status quo and to ensure that renewables become a normal part of your community.

The long-term goal of the UCS–NOAH collaboration was to make renewable energy and energy efficiency automatic components of NOAH's rehab work. This goal was achievable only with extensive follow-up support after completing the model home. First, we had to make sure the home was actually sold, and that the new owners recognized the unique value of their new home. UCS designed and printed a promotional flyer, in both English and Spanish, to help NOAH increase community interest in the home and its special features. Next, we organized on-site training for the new owners to explain the energy efficiency and renewable energy features and how to maintain them; the training was done by the energy services consultant who inspected the home. We wrote detailed homeowner's and renter's manuals to leave in the house, to help both the current and future inhabitants. We secured the ongoing help of an energy consultant to give NOAH technical assistance as they adapted the model house's energy-efficiency/ solar features to rental units. We also joined with other energy and environmental groups to exert pressure on the pending Public Utilities hearing to extend energy efficiency programs to multifamily rehab projects—the typical city housing stock.

## Conclusion

With persistence, the NOAH-UCS collaboration fulfilled its goals: UCS energy-efficiency recommendations have become "standard operating procedure" for NOAH's rehab work; NOAH will consider passive solar design on future rehab projects with a southern orientation; and two East Boston families are now living in a beautiful, comfortable home while paying modest utility bills. The project generated much interest in affordable housing circles, and we expect it to have an impact far beyond East Boston's borders. It was at once hard work and great fun. We surmounted many daunting obstacles and still suffered disappointing defeats. We learned a lot.

In the end, that's what this book is all about. It's about ordinary people who care about the future, who care about the environment, who care about energy, and are making things happen in their own neighborhoods and towns. This book provides many great examples of citizen-initiated ideas and plans to promote the use of renewable energy in their communities. These stories should give you ideas for projects in your own town or neighborhood.

But why should you bother? How can your town or neighborhood make even a ripple in the vast tide of energy consumption in this country? For far too long, U.S. energy policy has rested in the hands of so-called experts, and we Americans have not been well served. There has been scant progress at the national level, despite a supposedly sympathetic administration. Now, even the small gains that have been won are under severe attack from the "backlash" against environmental science in general, and climate science in particular. Concerns about climate change, environmental damage (like acid rain), and serious health impacts underlie all energy policy, and provide the most compelling rationales for significantly reducing fossil fuel use in our electricity generation and transportation.

What's needed is grassroots activism that will lead to successful renewable energy projects—like the ideas found in this book, and like the ideas that other inspired activists will soon create. It can be done; it must be done. We must take our energy future into our own hands and fashion it into the world we wish to leave for our children.

# Renewable Energy: How the Technologies Work

Don't be fooled by the modern-sounding term *renewable energy*. Humans have been using it in one form or another ever since we learned to bask in the sunshine and control fire. For eons we've depended upon firewood, dried manure, light and heat from the sun, the rush of flowing water, and the steady push of wind. Though today we call them by more scientific names—biomass, solar thermal, daylighting, hydropower—these ancient energy sources make even more sense today than do their modern successors—fossil fuels and nuclear fission.

*All* energy sources, with the exception of geothermal, tidal, and nuclear, ultimately originate with the sun. Winds are created by the uneven heating and cooling of the earth's surface. All green plants, from towering pines to microscopic algae, carry sunlight trapped in their cells. Since fossil fuels originated from decaying prehistoric plants and animals, they too owe their energy content to the sun. Rivers flow only as long as the sun evaporates water from oceans and lakes and generates rainfall.

As the term implies, renewable energy resources don't dwindle away with use; they will be around and available as long as the sun keeps shining. What's more, they can generally be used on the spot or close to where they are captured. This means that dollars spent on energy stay within a region or state and don't flow far from the local economy. Not so for fossil fuels, which most communities import from out-of-state or overseas, exporting cash in the process. Although use of any energy source affects the environment in one way or another, renewable resources generally have far smaller and more local impacts than do fossil fuels; in particular, they create less air pollution and its attendant health problems, and make almost no contribution to global warming.

According to the skewed accounting systems used in our society, most renewable energy technologies cost more than traditional fossil fuel technologies. That may be accurate when costs are measured only at the gas pump or the electrical socket. But add in the costs of environmental damage, of health problems due to air pollution, of the potential impacts of global warming, and of protecting oil supplies overseas, and renewable energy looks very attractive in comparison.

The United States is blessed with vast, inexhaustible renewable energy resources. In theory, North Dakota, Wyoming, and Montana alone could supply enough wind power to meet the country's annual electricity needs. Covering less than 1 percent of the country's land surface with solar collectors—that's one-tenth the amount of land used for agriculture—would generate enough energy to satisfy our annual energy demand. Our rich agricultural lands and thick forests could, if wisely managed, yield tremendous amounts of fuel for heat, electricity, and transportation, and raw material for products like glues and resins that are currently made from coal and petroleum.

Despite their common origins, different renewable energy sources demand different equipment and strategies for putting them to work in our homes, businesses, and vehicles. This appendix briefly describes basic technologies—their costs, advantages, and drawbacks. It also examines energy storage, an important component of many renewable energy systems. But first, let's take a look at a vital component of sustainable energy strategy: efficiency.

## The Importance of Energy Efficiency

In theory and in practice, renewable energy sources could surpass fossil fuel technologies as the predominant means of energy generation in the United States. But without a nationwide emphasis on energy efficiency, renewables will have to become more widely used just to maintain their present small share of the energy pie. *America's Energy Choices*, a study sponsored by four environmental organizations, including the Union of Concerned Scientists, makes this point very clearly. According to U.S. government forecasts, if we continue our current energy policies until 2030, energy consumption will rise 41 percent over 1988 levels, while renewable energy's share of the total will increase just 13 percent. If, however, we invest in aggressive energy efficiency programs and promote renewable energy technologies, we could actually reduce the nation's energy consumption by 18 percent while boosting renewables' contribution to 42 percent.

Many energy-efficiency measures are simply common sense applied to the design, construction, and maintenance of buildings, appliances, automobiles, and motors. If you orient a home so that it catches as much sunlight as possible during the winter, build it with plenty of insulation, add a tight vapor barrier, and make sure that doors and windows keep out drafts, then little extra heat is needed. In fact, owners of highly efficient homes sometimes can't find central heating systems small enough to meet their needs. New technologies, such as efficient lighting and motors, or improvements in industrial process efficiencies, such as freeze concentration technology (freezing water takes only 15 percent of the energy required for boiling it), offer comparable reductions in energy use in all sectors of the economy. But even the most efficient technologies cannot entirely eliminate the need for energy to power transportation, industries, businesses, and homes. As we become more efficient, we must also begin to replace some of our dependence upon fossil fuels with use of less harmful renewable resources.

## Wind Power

The steady turning of wind-powered water pumps helped make farming possible for settlers in the American Midwest and West. Ordered from the Sears and Roebuck catalog or available through the local hardware store, these twenty-blade machines used to be as common as barns and chicken coops on farms and ranches from Ohio to California. More than six million were sold before rural electrification put them out of business in the 1930s and 1940s.

Today's windmills bear little resemblance to their forebears. They have evolved into sophisticated machines which churn out electricity every bit as cleanly and much more quietly than their ancestors pumped water. More widespread use of these devices could supply a fifth or more of U.S. electricity needs, reducing coal use and its associated pollution by a similar fraction.

**HOW IT WORKS** Moving air can be put to work—children discover this when they blow on a pinwheel and set it spinning. To capture this energy and convert it into electricity, engineers and designers have built several types of wind machines, or wind turbines. Some look like huge fans, usually with three long blades. Others look like a giant's eggbeater, sitting straight up. No matter what shape the wind catcher takes, all turbines work essentially the same way. The blades

*The Darrieus wind turbine operates on a vertical axis. The 24-inch blades, shaped like a cross-section of an airplane wing, are capable of driving the generator regardless of wind direction. With the equipment at ground level, tower construction costs are reduced.*

Photo credit: U.S. Department of Energy.

connect to a single rotor which fits into the *nacelle,* or turbine housing. In most older machines, the rotor spins at a constant speed no matter how hard the wind blows, steadily turning a magnet through a coil of wire. This action generates a flow of uniform-frequency electricity.* Electronic equipment "conditions" the current before it is used to drive a pump or other device, or sent out on cables to the utility grid.

Today, wind machines come in two basic sizes: small and large. Small wind machines, built primarily for individual applications such as water pumping or home electricity production in areas far from utility electric supply, range from 0.25 kilowatts to 10 kilowatts. Bigger wind machines, generally used for utility-scale electricity generation, range from 50 kilowatts up to 750 kilowatts.

New and better wind turbines are being developed all the time. A current model marketed by Kenetech Windpower replaces the fixed-speed generator with a variable-speed generator and advanced electronics that can deliver a single-frequency current. This new design allows turbines to capture energy more efficiently over a wide range of wind speeds, and stand up to strong gusty winds better than their fixed-speed cousins. Other manufacturers are also marketing variable-speed machines.

---

*All turbines, whether they are spun by the wind, rushing water, or steam produced by boiling water, generate electricity the same way, with a magnet and a coil of wire. Rotation through the magnetic field causes a current to flow through the wire.

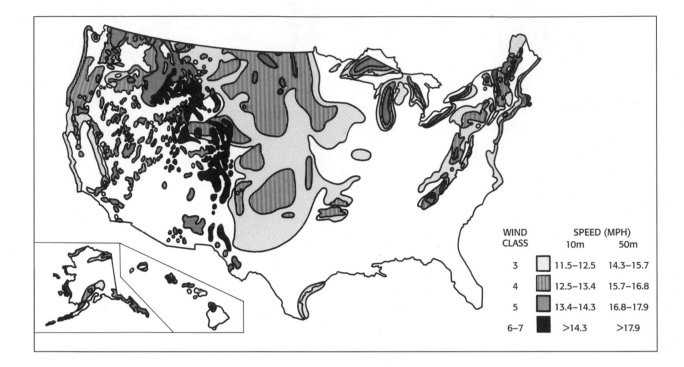

| WIND CLASS | | SPEED (MPH) | |
|---|---|---|---|
| | | 10m | 50m |
| 3 | | 11.5–12.5 | 14.3–15.7 |
| 4 | | 12.5–13.4 | 15.7–16.8 |
| 5 | | 13.4–14.3 | 16.8–17.9 |
| 6–7 | | >14.3 | >17.9 |

*Map of windy areas in the United States showing annual average wind speeds at two heights above the ground— 10m (33 feet) and 50m (164 feet). Class 5, 6, and 7 areas are generally suitable for development with today's technology; class 3 and 4 should become available soon with improvements in technology.*
Credit: Pacific Northwest Laboratory, 1986.

**RESOURCE**    Maps of average wind speeds reveal the obvious— some parts of this country are very windy and others aren't. But they also show that calm areas may sit right next to windy ones, thanks to the vagaries of topography. Thus, although the aggregate resource is potentially enormous, local availability is an important concern. Competing land uses may also determine the amount of wind potential that can be developed in a given area. In a detailed 1991 study by the Pacific Northwest Laboratory, researchers combined average wind speed and current land uses on a single map in an effort to calculate the practical potential of wind energy (see map, above). They calculated that, if advanced wind turbines can operate in so-called moderate wind areas (those with average wind speeds ranging from 11 to 14 miles per hour at a 30-foot height), then wind could deliver between two and six times the current U.S. demand for electricity. Much of this potential is in the Midwest, but most states (except in the Southeast) appear to have some excellent sites.

The largest wind projects to date are located in California, where, last year, more than sixteen thousand turbines generated approximately 2.85 billion kilowatt-hours of electricity. The American Wind Energy Association estimates that another one thousand (mostly small) turbines are scattered around the rest of the country, but their combined generating capacity is dwarfed by California's capacity. Wind development is beginning to pick up in other parts of the country, however. In Minnesota, Northern States Power has committed to a total of 425 megawatts of wind energy by 2002: 25 currently operating, 100 megawatts by 1996, 100 megawatts by 1998, and the remaining 200 megawatts by 2002. New England Electric has arranged for 20 megawatts of wind generating capacity to be built in Maine. All told, several hundred megawatts are scheduled for development in the next few years—a substantial boost to the U.S. wind industry, but still less than is planned in Europe.

## ADVANTAGES AND DISADVANTAGES

Wind offers a clean way to generate electricity using a locally available, free resource. Unlike coal, it produces no air pollution or solid wastes. Another advantage is that wind is a modular technology. In other words, a utility or business can start out with a small number of turbines, and add a few at a time when it needs more power or comes up with additional cash resources.

Yet wind potential is extraordinarily site-specific. Average wind speeds measured at a local airport won't necessarily match the average wind speed a mile away. Prior to planning a wind development, sites must be carefully surveyed and evaluated. Preferably, wind speeds should be measured at different heights and over the course of at least several months before embarking on a project.

Another drawback of wind power is that you can't order more wind when you need extra electricity or turn it off when you don't need any at all. Generation is totally dependent on when the wind blows, which requires some adjustment for our energy-on-demand society. For small systems, adding a backup source of energy such as a diesel generator or a battery storage system can provide a solution. Utilities generally get around this problem by using other power plants on their system to provide backup power. Although, kilowatt for kilowatt wind power does not provide as much reliable capacity as other types of power plants, it can nevertheless replace fossil

**THE TOP TWENTY STATES** for wind energy potential, as measured by annual energy potential in the billions of kWhs, factoring in environmental and land use exclusions for wind class sites of 3 and higher.

| | | |
|---|---|---|
| 1 | North Dakota | 1,210 |
| 2 | Texas | 1,190 |
| 3 | Kansas | 1,070 |
| 4 | South Dakota | 1,030 |
| 5 | Montana | 1,020 |
| 6 | Nebraska | 868 |
| 7 | Wyoming | 747 |
| 8 | Oklahoma | 725 |
| 9 | Minnesota | 657 |
| 10 | Iowa | 551 |
| 11 | Colorado | 481 |
| 12 | New Mexico | 435 |
| 13 | Idaho | 73 |
| 14 | Michigan | 65 |
| 15 | New York | 62 |
| 16 | Illinois | 61 |
| 17 | California | 59 |
| 18 | Wisconsin | 58 |
| 19 | Maine | 56 |
| 20 | Missouri | 52 |

*Several states have greater wind potential than California, where the vast majority of wind development has occurred to date.*

Source: *An Assessment of the Available Windy Land Area and Wind Energy Potential in the Contiguous United States,* Pacific Northwest Laboratory, 1991.

fuel capacity. As a rough rule of thumb, 3 or 4 kilowatts of wind will have the same capacity value as 1 kilowatt of conventional supply.

The whoosh of blades against the wind creates a low, steady drone. From a single well-built, well-maintained turbine it's almost inaudible. The noise from an entire wind farm, however, can't be missed. Some people also object to the strong visual impact of thousands of turbines turning on windy ridges. For these reasons, large wind farms will probably be relegated to areas away from cities and towns.

The steady winds that make for good windpower sites sometimes coincide with prime habitat for raptors—soaring birds like hawks and eagles—or stop-off points for migrating birds. Spinning turbine blades are hazardous. Bird casualties have turned out to be a major problem at one California wind farm where raptors live and hunt, and may become serious at another farm in southern Spain, where migratory birds stop to rest and wait for favorable winds before crossing the Mediterranean. Such problems must be avoided in the future through careful site evaluation and selection.

## Passive Solar Design

The simplest, cheapest way to tap renewable energy seems to get the least attention. Passive solar design involves building homes, offices, and other buildings so they take advantage of reliable daytime light and heat. Without using any machines or devices other than glass, insulation, and heat-absorbing walls and floors, passive solar design can provide a building with a large fraction of its heat and daytime light.

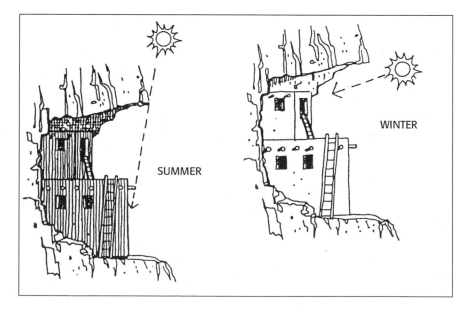

*This diagram illustrates how cliff dwellers in the Southwest maximized the use of passive solar design for their living space. The cliff's overhang shields dwellings from the fierce summer sun at its highest arc, while the sun's winter angle pours solar energy into the cliffs, providing warmth and light.*
Credit: PPG Industries, Inc.

Passive solar design was once so important it was mentioned in the Justinian Code, a collection of Roman laws compiled in the sixth century. The Code protected sunrooms from being shaded by a neighbor's building, trees, or other shadow-casting objects. (Solar covenants recently established in cities like Portland, Oregon, and San Jose, California, protect modern solar homes in a similar fashion.) Cliff dwellers in the American Southwest built their homes into south-facing cliffs, where they would get direct sunshine during the winter and be shaded from the sun during the summer.

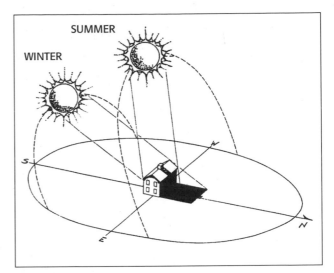

**HOW IT WORKS** Several architectural tricks can dramatically reduce a building's energy consumption, even during a snowy winter in the Northeast or a boiling Texas summer. The basic strategy requires orienting a new home so its long axis runs east-west, and placing most of the windows on the south-facing side. This orientation guarantees that the low winter sun will shine into the building during most of the day. During the summer, when the sun traces a higher arc, overhangs keep direct sunlight out. In some passive solar buildings, winter sunlight beams right into the living space and falls on a heat-absorbing wall or floor. Generally made of masonry or tile, the wall or floor slowly releases its stored heat well into the night. Indoor ponds or beds of crushed rock can also store up the day's sun-generated heat. In other passive designs, a special sunspace intercepts sunlight, heats up, and then distributes warm air through vents to the rest of the building.

The north side of a passive solar building has few windows, and may be built into a hill to keep heat loss to a minimum. Plenty of insulation and tight fitting double- or triple-paned windows that trap heat inside during the winter or keep it out during the summer are key requirements of passive solar design. After all, it doesn't make much sense to let free heat from the winter sun leak right out, or to let the summer sun turn a building into a furnace.

**COST** The real beauty of passive solar design is that it can be implemented with little or no extra cost. An experienced solar architect can orient the home or building with its long face to within at least 30 degrees of true south, creating energy savings without even changing

*The relationship between the sun and the earth results in a summer sun high in the sky and a winter sun lower in the sky. The placement of the sun in the sky at various times of year has a direct bearing on the location and construction of any solar project, and unobstructed access to the sun's energy must be protected to take advantage of passive solar design.*
Credit: Vermont Energy Office.

the design. The expense of adding more windows on the home's south side is offset by having fewer windows on the north side. Even shading devices on the south-facing windows to prevent summertime overheating add only a couple thousand dollars to construction costs.

**ADVANTAGES**   Passive solar design is a one-time expense that provides free heat and light for as long as the building lasts. Little maintenance is needed to keep this energy system operating in peak condition. If a passive solar home is built with south-facing windows averaging 7 percent of the house's total floor space, there are no extra construction costs, and the homeowners can save up to 25 percent of the house's conventional heating fuel. Neuffer Construction of Reno, Nevada, has developed passive solar tract home designs that add only 1 percent to the cost of construction while delivering a more than 50 percent annual energy savings! Even adding passive solar to gut rehabilitation of existing structures can prove cost-effective if the building's orientation is right. It doesn't take long for the owner's initial investment to be repaid several times over.

Passive solar design partially replaces oil, natural gas, or electric heat with a free, nonpolluting, quiet, local, and renewable energy source. Just as important, passive solar homes are bright and airy; people who live in them appreciate the spacious and attractive atmosphere they create. They can be built anywhere there's access to the sun.

## Solar Thermal

Many hikers and campers know the value of a black, watertight plastic bag filled with water and hung in the sun—there's nothing like a solar-powered hot shower at the end of a strenuous hike. This simple use of sunlight for heating water isn't a new concept. In the 1930s, solar systems supplied hot water for more than half the homes in Miami. French inventors in the seventeenth and eighteenth centuries developed solar collectors capable of generating temperatures well above 1,500 degrees C, and built engines powered by solar-generated steam.

In the United States today, solar thermal energy could help meet two divergent needs: hot water for homes, schools, and offices, and electricity for utilities.* Unfortunately, even at the market's peak in the early

---

*In some developing countries, people spend hours each day searching for dry wood for cooking. Solar ovens could reduce this daily hunt, and save dwindling supplies of firewood. Small solar thermal electricity generators could also supply electricity to isolated villages.

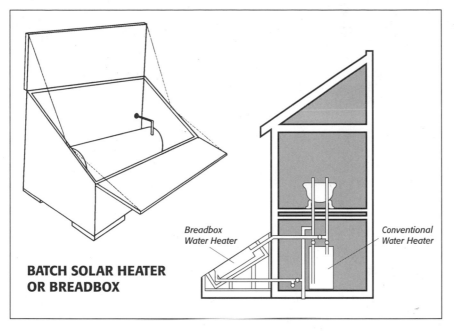

**BATCH SOLAR HEATER OR BREADBOX**

*Breadbox Water Heater*

*Conventional Water Heater*

*A batch solar heater is a simple passive system that works well as a water preheater. It has no moving parts.*

Credit: *Renewable Energy Native to California*, California Energy Extension Service, 1993.

1980s, solar collectors did not play a significant role in the nation's energy supply, and were used primarily for heating hot tubs and swimming pools. This doesn't have to be the case: in Cyprus, for example, more than 90 percent of homes are now equipped with solar water heaters, as are at least 65 percent of homes in Israel; in parts of Australia, the market penetration is above 30 percent. Even a modest increase in U.S. sales would bring down the cost of solar water heaters, since manufacturers would be assured of a more secure market and could then take advantage of economies of scale.

Solar heat can also be used to generate electricity. Beginning in 1984, an Israeli-based development company, LUZ International, generated electricity using solar thermal energy in the California desert, and sold it to a utility. Although the company went bankrupt, the solar thermal power plants are still operating. And several of California's largest utilities are testing different, improved solar thermal concepts.

**HOW IT WORKS** When sunlight is absorbed by a material, it turns into heat. Solar thermal systems put this principle to work heating water for homes, businesses, and swimming pools. The basic equipment for doing this hasn't changed much since the first collectors were built and sold in the 1890s. A standard solar water heater consists of a shallow, well-insulated glass-topped box. A black metal plate made of copper or aluminum lines the inside, absorbing sunlight and radiating heat. Snaking back and forth inside the box, a long copper tube

*This large-scale solar thermal plant at Kramer Junction, California is one of only a few such commercial operations in the United States, selling solar-generated electricity to the area's utility. With the additional capability to use natural gas as back-up for those occasional cloudy periods, this plant ensures the utility a steady source of electricity.*
Photo credit: Luz International.

filled with water or an antifreeze mixture absorbs this heat. As water cycles from the collector to a storage tank and back again, its temperature can rise to over 100 degrees C (the boiling point of water, equivalent to 212 degrees F). Approximately 3 to 4 square meters of collector space are required to provide hot water for a family of four; the size of the system is governed by climate and needs of the user.

Active solar water heaters use an electric pump to push water up to the collector. A photovoltaic cell is ideal for providing this power, since both operate only when the sun is shining. Passive systems rely on convection or the bubbles formed by boiling water to drive this flow.

Solar thermal electricity generating plants incorporate strategies for concentrating the sun's heat. The LUZ plants in the Mojave Desert use long rows of curved mirrors that concentrate sunlight onto a liquid-filled pipe running along their center. This hot fluid is used to boil water, and the resulting steam drives a turbine. Solar Two, a "power tower" currently under development, takes a different approach. It focuses the sunlight reflected from 435 mirrors onto an opening at the top of a 300-foot tower. Here the brilliant light falls on exposed pipes carrying melted sodium, heating it well above 1,000 degrees C. This molten mass is then used to boil water and drive a steam turbine.

**RESOURCE AND COST**   Unlike the case of wind energy, the application of solar thermal energy in the U.S. is not really constrained by the size or distribution of the resource. Every part of the country receives sunlight, and there is no more than a factor of two difference between the amount of sunlight falling on the sunny Southwest and the cloudy Northeast on an average basis. A more important constraint is cost. An active solar water heating system for a family of four, installed, costs between $2,000 and $4,000, depending on type, materials, installation, and service. A passive system such as the one-size-fits-all Copper Cricket™ costs around $2,500 installed. While substantially more expensive than a comparably sized electric water heater, monthly fuel savings can pay back the incremental costs within a few years. Natural gas remains less expensive, however.

**ADVANTAGES AND DISADVANTAGES**    A home-sized solar water heater sitting unobtrusively on a roof provides hot water without using fossil fuels. That means it can offset substantial amounts of carbon dioxide, nitrogen oxide, sulfur dioxide, and particulate emissions, without any fuel costs. Solar thermal systems come in a variety of sizes to meet different needs, and can be placed almost anywhere—on the roof, in a backyard, atop a garage. A Minnesota company is even making pool heaters that use a home's existing roof as a solar collector. The waste heat that builds up in the attic is used to heat pool water in a cycle which also helps cool the house.

Unfortunately, it's just not possible to run the sun longer whenever guests arrive for an extended visit and you need extra hot water. And cloudy stretches pose a problem. That's why most people who install solar water heaters also retain a backup system using electricity or natural gas. Solar thermal electricity generating plants also use backup gas or storage systems to smooth out the sun's variability, enabling these plants to produce electricity with no air pollution and only fuel costs associated with backup needs. With backup, solar thermal systems are as reliable as conventional electricity generating facilities.

## Photovoltaics (PVs)

Unlike wind and solar thermal energy, turning sunlight directly into electricity with solar cells is a relatively new technology, a product of the Space Race. In 1954, scientists at Bell Laboratories in Princeton, New Jersey, discovered they could use crystals of silicon that were "doped," or laced, with small quantities of other elements to convert sunlight directly into electricity. Within several years, solar cells were generating electricity for satellites launched into space. Research has since yielded a variety of these photovoltaic cells, for use in space or on earth, some of which can convert sunlight to electricity at an efficiency of over 30 percent.

**HOW IT WORKS**    Just as a sheet of water is made of billions of tiny droplets, a ray of light is composed of billions and billions of photons—sub-sub-microscopic packets of light energy. When a photon of light strikes the silicon and trace-element surface of a photovoltaic cell, it knocks a negatively charged electron out of its orbit around a

positively charged nucleus. Normally, this electron would quickly make its way back to the nucleus. But in a photovoltaic cell, the junction of two different "doped" layers sets up a voltage that pulls the electron in one direction and the positively charged "hole" in the other. Close the circuit with contacts and wires and, voilà, an electric current flows. Single photovoltaic cells generate between 0.6 and 1.2 volts; several are usually wired together to yield higher currents and voltages.

Silicon remains the most common material used in photovoltaic cells, though other substances such as gallium arsenide may also be used to improve the efficiency of conversion of light to electricity. In one approach, the doped silicon is grown into long crystals that must be sliced into thin wafers, almost like cutting bologna. Not only is this a painstakingly slow process, but much of the expensive crystal ends up being wasted. Another approach involves depositing noncrystalline silicon onto ultrathin films. Solar cells using this "amorphous" silicon are easier and cheaper to make, but are much less efficient than those made from single crystals. Other types of photovoltaic cells include thin films made of copper, indium and selenium, or cadmium and tellurium; different types of cells may also be stacked together to boost overall efficiency.

Solar cells now pop up in a variety of common products. Tiny ones power wristwatches and calculators. Catalogs offer porch and walkway lights that get all their juice from a few photovoltaic cells and a small battery. Homeowners who live far from the utility grid can buy photovoltaic arrays for the roof or backyard that generate 1,000 watt-hours a day, enough electricity to run most modern appliances. To be most useful, the devices run by PV cells must be highly efficient and should run on direct current (DC).

Photovoltaic systems are also attracting utilities' attention. The Sacramento (California) Municipal Utility District currently operates the world's largest PV power plant, a 2-megawatt system built in 1986 right next to the closed Rancho Seco nuclear power plant.* New England Electric Company is investigating a different way to use photovoltaics. It has outfitted thirty homes, a Burger King restaurant, and several other businesses with rooftop PV panels. Electricity generated during the day is used in the home or business, and any left over goes into the utility grid, turning the customer's meter backward (a practice known as *net metering*). At night, the buildings draw power from the grid, which acts like a huge battery backup system.

**COST**    Like solar thermal energy, PV applications are limited not by the size of the resource but by cost. Since the first photovoltaic cells accompanied satellites into space in the late 1950s, their cost has dropped dramatically. The earliest units cost $44,000 per peak kilowatt of output. Today's commercial systems cost anywhere from $6,000 to $15,000 per peak kilowatt, depending on size and type. For example, a 200-watt system generating an average of 750 watt-hours per day, batteries and other equipment included, costs under $3,000.** Although that sounds like a lot of money for electricity, it may be far less expensive than paying the utility company to extend its distribution lines. Factor in the free electricity you'll receive for fifteen or twenty years and $3,000 may represent a much better deal than conventional grid power.

It is expected that the cost of photovoltaic modules will drop to between $3,000 and $5,000 per peak kilowatt in the next several years, and ultimately may reach $1,000 per peak kilowatt. This price drop won't happen all by itself, though. Improving cell efficiencies will help, since fewer square feet of expensive material will be needed to generate electricity. Expanding the number and types of applications appropriate for photovoltaics will help even more by increasing orders for the few existing solar cell makers. Bigger orders mean less overhead per unit made and greater automation in manufacturing.

---

*The largest PV plant ever built was the 6.4-megawatt Carissa Plains plant in California. It operated for several years, then was dismantled and its collectors sold.

**Price from the *Solar Living Sourcebook,* Eighth Edition, John Schaeffer and the Real Goods Staff (White River Junction, Vermont: Chelsea Green Publishing Company, 1994).

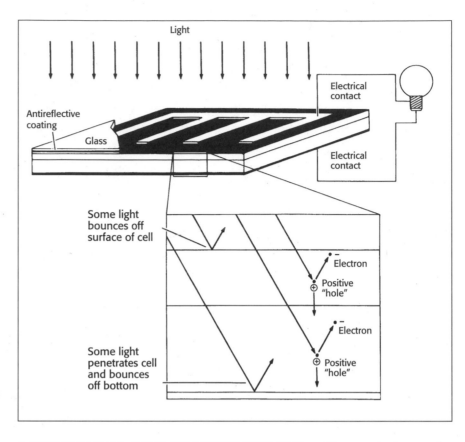

*Schematic of a typical photovoltaic cell. An electric field is created at the junction of two dissimilar semiconductor materials. When electrons are knocked loose from atoms by incident light, the negative electrons and positive "holes" are swept onto opposing electrical contacts. When the circuit is completed, a current is established.*

Credit: Union of Concerned Scientists.

Within the figure:

Light

Electrical contact

Antireflective coating

Glass

Electrical contact

Some light bounces off surface of cell

Electron

Positive "hole"

Electron

Positive "hole"

Some light penetrates cell and bounces off bottom

**ADVANTAGES AND DISADVANTAGES** Photovoltaics are the quintessential modular energy supply. You can start with one or two arrays and add extras when you can afford them, or as you need more electricity. They work soundlessly, emit no pollutants, require no water, and have no moving parts to maintain. Tests show that photovoltaic panels can be easily and safely integrated with the electrical grid, providing clean, renewable electricity for homeowners and helping utilities reduce their peak demand during the daytime.

Photovoltaics, like solar thermal systems, operate only when the sun shines. Sunshine and demand don't always coincide, making some sort of storage or backup a necessity in most applications. For remote homes far from utility lines, a backup system usually entails batteries. For those connected to a utility grid, the grid itself serves as backup. Utilities investing in large-scale photovoltaics may use a combination of the sun and other energy sources to meet peak daytime demand.

Some critics point out that photovoltaic plants capable of generating megawatts of electricity must spread out over large areas and thus pose serious land-use constraints. This could indeed be a problem if we move to large-scale solar electric generation, for example, as part of a solar hydrogen economy. For the present, however, creative solutions can minimize land-use concerns. In Austin, Texas, 300 kilowatts of photovoltaic collectors sit atop the 3M Corporation's Austin center, generating electricity for the center's parking garage. In Switzerland, photovoltaic panels line several highways and rail lines on otherwise unused and unusable land.

## Biomass

There's far more to biomass as an energy source than the hot crackle of dry wood in a fireplace or wood stove. This technology includes all plant material—trees, wheat stalks, grasses, even peanut shells. It also includes human and animal wastes, as well as municipal garbage. Some companies generate electricity by burning wood chips or crop residues. "Waste" wood now provides the paper and lumber industries with most of the heat and electricity they need. Wood, grasses, and other so-called energy crops could also provide us with ethanol for use as a transportation fuel, or fuel for power plants. And gas released by decaying garbage in landfills across the country is being trapped and burned to generate electricity.

*The biomass boiler heats water, which creates steam, which powers the turbine, which turns the generator, which generates electricity. Water from the turbine goes into a condenser to be cooled before it's returned to the boiler. Ash byproducts and smoke go up in the electrostatic precipitator for clean-up before emission.*

Credit: *Renewable Energy Native to California,* California Energy Extension Service, 1993.

**HOW IT WORKS**    All forms of biomass contain sunlight stored in chemical bonds. Plants use the energy in sunlight to transform carbon dioxide and water into sugar and other complex organic compounds. The bonds between adjacent carbon, oxygen, and hydrogen

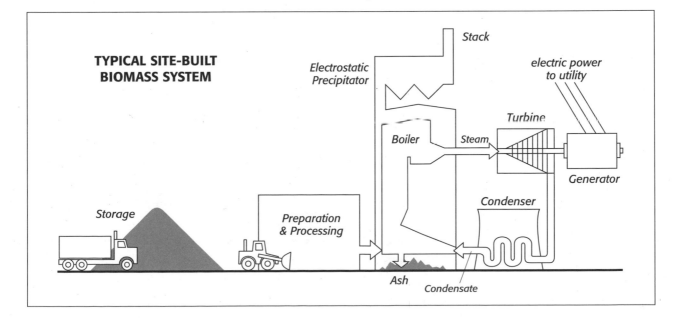

TYPICAL SITE-BUILT BIOMASS SYSTEM

Storage
Preparation & Processing
Electrostatic Precipitator
Stack
Boiler
Steam
Turbine
Generator
electric power to utility
Condenser
Ash
Condensate

atoms store energy (think of a compressed spring). Breaking the bonds by digesting, burning, or decomposing these substances releases the stored energy.

Simple wood stoves currently help heat millions of homes from New Mexico to Maine. Old or antique models tend to be very inefficient, sometimes actually losing heat overall because cool air is sucked into the home to feed the fire. Modern stoves, by comparison, can be highly efficient, especially if the heated air is circulated though masonry walls before escaping the chimney. In forested parts of the country, heating with wood may cost less than using gas, oil, or electricity for heat.

Large-scale biomass conversion—for heating a school or factory, or generating electricity—is sometimes done using a mammoth version of the wood-burning stove. New technologies simplify the process and make it more efficient. One company is testing boilers that burn whole trees, which would eliminate a costly investment in wood chippers. Other systems heat wood chips or other dried biomass in the absence of oxygen, cooking out flammable gases, which are then completely burned in a separate chamber. These gases can also be collected, purified, and liquified, making renewable substitutes for petroleum products such as gasoline and diesel fuel.

Municipal solid waste, or garbage, is about 60 percent biomass—paper, yard waste, kitchen scraps, and other plant- or animal-derived material. Some cities funnel their garbage into waste-to-energy plants. The heat it gives off when burnt is used to boil water, and the steam used to turn turbines to generate electricity. Some of these plants burn raw trash, others convert it into dried pellets called *refuse derived fuel*. In a similar fashion, dried sludge from sewage treatment plants can also be incinerated for energy.

What may be garbage to us is a veritable feast for bacteria. In the moist darkness of a landfill, a boggling array of microbe species slowly digest organic matter and leave behind their own waste, methane gas. More than 110 communities around the country currently tap their landfills to recover this energy-rich gas. It can be sold to businesses or burned to generate electricity. Such landfill power stations currently account for more than 315 megawatts of generating capacity.

**RESOURCE**     Until the mid-1800s, wood supplied 90 percent of the United States' energy needs. Then coal and, more recently, oil and natural gas pushed biomass into the background. In 1992, biomass accounted for just 4 percent of primary energy consumption. With careful management, however, biomass could again play a more visible role.

Natural and managed forests supply virtually all the wood, bark, and logging leftovers (tree stumps, gnarled trunks, branches, and leaves) used for energy in the United States. A promising new approach involves growing energy crops on idle or unused cropland. Research coordinated by the Oak Ridge National Laboratory in Tennessee has identified several plants that might make excellent candidates. These include trees—hybrid poplar, black locust, sugar maple, and eucalyptus—as well as fast-growing plants like switchgrass. One estimate* suggests that energy crops grown on idle and marginal land could supply roughly one-quarter of our current energy needs, while forestry products and waste could supply another 17 percent.

*Dedicated energy crops—like this chopped switchgrass—can be grown on idle or unused cropland, potentially providing a sound economic basis for farmers to allocate some land to high-yield, high-energy value biomass.*
Photo credit: Eric Denzler.

**COST**   Large-scale wood or biomass heating systems require considerable start-up expenses. A biomass gasifier capable of heating a 300-student high school in Vermont cost $265,000 in 1992, plus another $200,000 for stringing a mile of hot-water–carrying copper pipes from room to room. That hefty initial investment will be offset by substantially lower fuel costs, compared to the electric heat that had been used for the previous twenty years. Switching to locally grown wood will cut the school's annual fuel bills by 90 percent, and save the district more than $1 million over a twenty-year period (see Chapter 4).

**ADVANTAGES AND DISADVANTAGES**   Biomass is a versatile, local energy resource. Those living in forested areas can use wood or wood chips; in agricultural areas, crop wastes such as corncobs or peanut shells; near cities, pelletized trash makes a usable energy resource. Imaginative uses for biomass abound. In California each year, several small power plants consume more than a quarter of a million tons of construction wood waste, discarded plywood, and landscaping refuse that would otherwise wind up buried in landfills. Biomass can also be

---

*This estimate is from an article by Cook, Beyea, and Keeler, "Potential Impacts of Biomass Production in the United States on Biological Diversity," *Annual Review of Energy and the Environment,* Annual Reviews, Inc. (Palo Alto, California, 1991), 401.

made into transportation fuels. Corn and even wood can be fermented into alcohol for fuel, while certain seeds such as soy beans and rape-seed produce a diesel-like fuel oil. Increasing the use of biomass for energy could provide new and continuing jobs for farmers, foresters, and loggers around the country.

Biomass may, however, be the trickiest renewable energy resource to manage properly. Our familiarity with it, combined with its versatility, make biomass easier to sell and put into use than, say, wind power. If it is not managed carefully, though, increasing biomass' contribution to our energy mix could damage forests, erode cropland, and do little to ease air pollution.

Burning wood, alcohol, or dried sewage generates essentially the same set of air pollutants that come from burning coal or natural gas. Cities with a large number of wood-burning homes like Albuquerque, New Mexico, have actually had to ban wood-burning during certain winter days in order to minimize the sooty haze that hangs over the city. All modern stoves must now meet stringent emissions guidelines set by the EPA so they release little in the way of soot or other pollutants, but many older, polluting home stoves are still in use. Large bio-

## CUTTING-EDGE BIOMASS TECHNOLOGIES

**Whole-tree combustion.** Whole-tree combustion is a modification of currently existing direct-combustion technologies, and uses whole trees rather than chipped wood, avoiding the expense of the chipping process. Waste heat from the power plant is used to dry the wood before combustion, and the steam system is operated at higher temperatures and pressures than current biomass power facilities, improving the overall efficiency of the process.

**Biomass gasification/gas turbine.** When biomass is heated in an enclosed reactor vessel, the solid biomass volatilizes into a combustible gas. The gas can be burned in a standard steam turbine system, or cleaned and used as fuel in a gas turbine in place of natural gas. Gas tur-

bine systems are potentially more efficient than steam turbines, especially when part of a combined cycle (gas turbine and steam turbine) system. Several scale-up demonstrations of gasification technology and related systems technology are currently taking place.

**Fuel cells.** A fuel cell is an electrochemical technology that generates electricity when hydrogen and oxygen combine to form water inside the fuel cell; no combustion is involved. Gasified biomass, or alcohol fuels derived from biomass, could be used to provide the hydrogen needed to generate electricity in fuel cells. Fuel cells are also being developed to use methane from landfills or agricultural waste operations as their feedstock.

From *Biomass Power Commercialization: The Federal Role,* A Report from the Union of Concerned Scientists (May 1994).

mass burners must be fitted with scrubbers to ensure they don't emit excessive pollutants. This is especially true for waste-to-energy plants, since their emissions may contain toxic metals or dioxins formed by burning plastics.

Responsible land use presents another tricky issue. Increasing use of wood as an energy source will further boost our demand for trees, and could have a potentially devastating impact on our already fragile forests. Overharvesting and clear-cutting cause erosion and the silting of streams, and destroy habitats for a variety of plants and animals. The spotted owl/old growth controversy that is still simmering in the Northwest illustrates how complex and fiery the issues surrounding forest resources can be. Potential problems even surround proposals to grow certain energy crops. Using annual row crops like sweet sorghum to make biofuels could exacerbate soil erosion, and increase the use of agricultural chemicals.

Most of these problems can be managed, however, so long as care is taken to ensure sustainable biomass use. For instance, the use of perennial trees and grasses for energy can reduce erosion on vulnerable soils, since they do not require annual plowing and planting. And current forest practices can be greatly improved. But all of this requires appropriate regulation, which is not yet in place.

## Hydropower

If you've ever tried to wade across a fast-flowing stream, or stood under a waterfall and let the water pelt your head and shoulders, you've felt the inexorable power of moving water. The ancient Greeks built simple waterwheels to harness this power for grinding grain. Early New England colonists built more than ten thousand waterwheels before 1800. By 1878, George Westinghouse had built the first of several turbines at Niagara Falls. He converted their output into alternating current, a step that made long-distance transmission of electricity practical. Little has changed since then, save for small refinements in equipment and the size of hydropower projects.

**HOW IT WORKS**   Water power is yet one more example of stored sunlight. When the sun's energy beats down on oceans or lakes, water evaporates from the surface and rises into the atmosphere. When this moist air rises over a mountain or hits a front of cool air, it condenses and falls as rain or snow. From the moment it touches down, water trickles, creeps, flows, and rushes downhill and downstream, eventu-

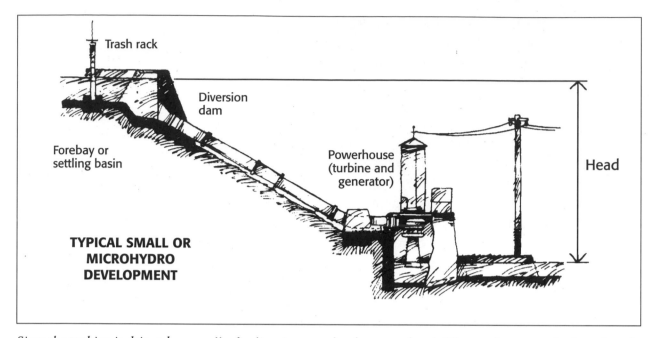

**TYPICAL SMALL OR MICROHYDRO DEVELOPMENT**

Trash rack

Diversion dam

Forebay or settling basin

Powerhouse (turbine and generator)

Head

*Since the turbine is driven by moving water in a typical small hydro development, it is critical to have sufficient head to create the force needed for the water to turn the turbine.*

Credit: *Guide to Development of Small Hydroelectric and Microhydroelectric Projects in North Carolina,* Energy Division, North Carolina Department of Commerce.

ally finding its way back to sea level. The sun's energy is therefore *in* the flowing water.

Several types of machine have been invented to capture the energy stored in running water. Run-of-the-river turbines sit submerged in a river or stream and turn as water rushes through them. The spinning turbines turn magnets through coils of wire, generating a current of electricity. Such turbines don't need a dam, thus keeping costs down and minimizing the environmental impact. On the other hand, they are susceptible to variations in river flow, and can't operate when the water drops below a certain level.

The more commonly used hydropower systems harness the energy in *falling* water. They generally use a dam to create an abrupt drop, called the *head*. The vertical distance between the water behind the dam and the free water downstream determines how much energy can be captured. The bigger the drop, the more energy is available. Dams also store water, smoothing out daily and seasonal changes in river flows. Water in front of the dam is diverted into one or more downward-angled pipes called *penstocks*. It falls through these pipes and turns turbines placed at the bottom. The slower, less energetic water emerges from the turbine and continues downstream. If a river is big enough, and drops substantially in elevation from source to mouth, it can support several dams and hydropower stations. The Columbia River in Washington and Oregon, for example, supports seven dams and hydropower stations.

When we hear the word "hydropower," we tend to think big. But even small streams can be used to generate electricity, as long as they flow fast enough and a big enough head can be created. In fact, several companies sell kilowatt-scale hydropower systems for homeowners living in remote areas.

Water power also comes in three less commonly used forms: tidal power, wave power, and ocean thermal energy conversion. The ultimate potential of these technologies to contribute to U.S. and international power generation remains to be seen. *Tidal power* combines the basic principles of river-based hydroelectric generation with the huge tides that occur in certain parts of the world. Massive gates close immediately after the incoming tide hits its highest point, creating a temporary tidal bay. Just before low tide, when the head is at its maximum, water is released from the dam through a series of turbines. Once it has spilled out, the gates reopen and the cycle starts again. The oldest and largest such plant is a 240-megawatt station at LaRance in northwest France. An experimental 20-megawatt tidal plant is located in Nova Scotia's Bay of Fundy.

*Wave power* captures the ceaseless up-and-down motion of sea water and converts it into electricity. It works quite well for powering buoy markers and navigational warning beacons. Larger systems have been proposed and a few tested, but none yet appear to be feasible.

*Ocean thermal energy conversion* uses heat collected in the top layer of ocean water to boil an easily vaporized substance, such as ammonia. The resulting vapor turns a turbine. Icy cold water drawn from the ocean's depths cools and condenses the vapor. Ocean thermal energy plants have successfully generated electricity twice, first off the coast of Cuba in 1929, and again off Hawaii in the early 1980s. Making this technology economically feasible, however, is proving difficult.

**RESOURCE**   Rivers currently contribute somewhere between 10 percent and 12 percent of the U.S. electricity supply, depending on rain and snowfall. That's the equivalent of fifty nuclear power plants. There is some potential for expansion, but not nearly as much as for other renewable resources. Dams and hydropower plants have been constructed in the United States on a regular basis for the last seventy or eighty years, making hydro a mature renewable energy source—most of the best sites have been developed and much of the potential has been tapped. While the profound environmental impacts of mega-projects such as Hoover and Grande Coulee dams make these unattractive options, opportunities still exist to develop smaller hydropower plants. For example, Cornell University recently renovated a 1.2-megawatt plant near its campus that had been shut down in the late 1960s. The university gets approximately 5 percent of its electricity needs from the plant, and sells the local utility about $250,000 worth of electricity each year.

**COST**     Traditionally, hydropower has been one of the least expensive sources of electricity. Some of the immense facilities that were built in the heyday of hydropower construction generate power today at a cost of less than one cent per kilowatt-hour. The cost of new hydropower plants varies widely, and depends on such factors as the size of the plant, the design, the capacity factor, licensing delays, environmental mitigation costs, and proximity to transmission lines. As is the case with most renewable energy sources, initial capital costs are relatively high, whereas operations and maintenance costs are low. Uprating the capacity of an existing plant costs much less than building a new plant.

**ADVANTAGES AND DISADVANTAGES**     Water power represents a clean source of electricity, one that is virtually free once the equipment costs needed to harness it have been paid back. Also, since most hydropower plants have built-in storage capacity—the water behind the dam—they make good companions to solar or wind power, which are limited by periods of inactivity. Hydropower systems come in all sizes, from 10-kilowatt modules just right for a stream-side home, to several hundred kilowatt systems sufficient for powering a business, to a multi-megawatt project that can generate enough electricity to light up a small country.

The main constraints on hydropower development are its impacts on fish, wildlife, and local communities. Though hydropower plants generate electricity without polluting the air, dams flood large areas, covering much more land, in fact, than solar facilities producing the same amount of power. Turning a narrow stretch of moving water into a wide, still pond or lake completely alters the ecosystem, disrupting long-established plant and animal communities and eliminating the scenic beauty of a tumbling open river. A variety of federal, state, and local regulations governing hydropower development reflects these concerns. Thus, building even a small hydropower station requires patience, close work with local citizens and environmental groups, and constant interaction with government agencies.

All the energy we could ever use sits just below our feet if—and this is a pretty big if—we can find inexpensive, efficient, and clean ways to extract it. Geothermal energy, which takes several different forms, theoretically dwarfs the United States' proven reserves of coal, our largest domestic fossil fuel resource. But extracting even a fraction of this energy will demand creative research and engineering, and substantially increased government funding.

Geothermal energy is the only major renewable resource that isn't derived from the sun. Instead, the radioactive decay of unstable elements like thorium and uranium in the earth's core releases heat, which remains trapped deep underground. Thanks to deep faults and fissures, though, groundwater can sometimes percolate down to these hot rocks. As the water heats up, it rises back toward the surface, where it can be captured for use.

The Greeks and Romans built public baths over hot springs, or piped in this naturally warmed water. Native Americans in California and Guatemala used hot springs and steam vents, called *fumaroles,* for cooking. Early settlers in Klamath Falls, Oregon, and Boise, Idaho, tapped shallow pockets of hot groundwater and heated their homes with it. Their descendants continue the practice today. And since 1960, a huge reservoir of geothermal steam and other hot-water reservoirs in the West have been tapped to generate 3,000 megawatts of electricity.

*Ground source heat pumps* are often considered to be a form of geothermal energy. That's not completely accurate. These devices require a medium with a constant temperature to act as a source or "sink" for heat pumped into or out of a building. The medium could be cold well water or water running through city pipes or the ground below the frost line. They also require electricity to pump heat from this source to a home. Even though heat pumps don't qualify as a renewable energy technology, they make some use of geothermal energy, and are so efficient they deserve mention here.

**HOW IT WORKS**   Geothermal resources come in four basic forms: hydrothermal, hot dry rock, magma, and geopressured brines. All but hydrothermal are still very much in the experimental phase and, if proven feasible, may only be suitable for utility-scale development, not for independent home or business use.

*Hydrothermal* reservoirs come in two basic types, those dominated by hot water and those containing mostly dry steam. The latter are very rare and highly prized, since the pressurized steam that rushes

*Geothermal*

out from drilled wells can turn a turbine and directly generate electricity. Hot water from geothermal reservoirs can be used for heating homes, drying food or paper, pasteurizing milk, and generating electricity. In Reykjavik, Iceland, and Elko, Nevada, geothermal hot water is pumped to homes just like drinking water. It is circulated through floors and radiators, providing clean, reliable heat.

*Energy is created by super-heated water from deep beneath the Earth's surface.*

Graphic credit: Geothermal in California, Department of Conservation, Division of Oil, Gas, and Geothermal Resources.

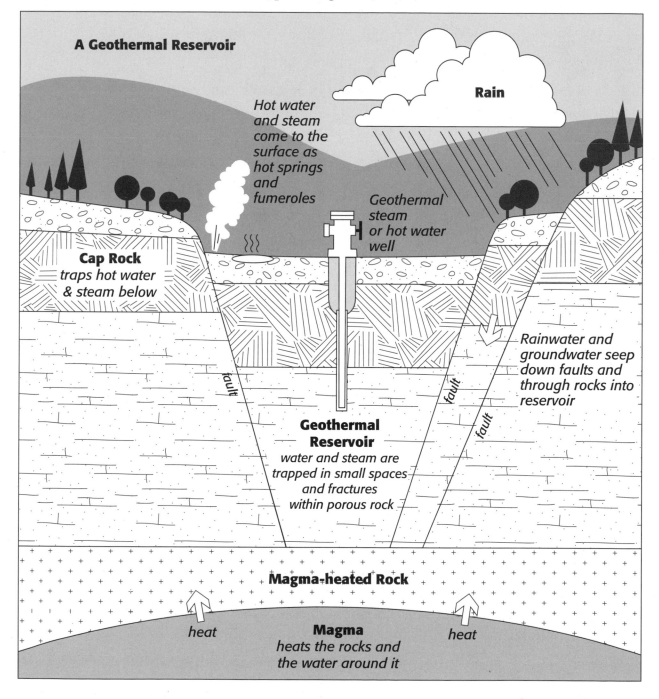

**A Geothermal Reservoir**

Rain

*Hot water and steam come to the surface as hot springs and fumeroles*

*Geothermal steam or hot water well*

**Cap Rock**
*traps hot water & steam below*

fault

fault

fault

*Rainwater and groundwater seep down faults and through rocks into reservoir*

**Geothermal Reservoir**
*water and steam are trapped in small spaces and fractures within porous rock*

**Magma-heated Rock**

*heat*

*heat*

**Magma**
*heats the rocks and the water around it*

If the intriguing concept of *hot dry rock* ever pans out, it could bring geothermal energy to all parts of the country, instead of just to those areas lucky enough to sit atop hydrothermal reservoirs. Research into hot dry rock technology is based on the fact that rock temperatures rise the further you drill into the earth. On average, they climb about 40 degrees C per mile, though in certain parts of the West this temperature gradient exceeds 100 degrees C per mile. Tapping into this vast energy source involves drilling down into a layer of hot rock, fracturing it with hydraulic pressure, then pumping down a steady stream of water. The fractured rock acts like a huge heat exchanger, transferring the earth's heat to the water. If the rock is hot enough, it could heat water close to or above its boiling point, making electricity generation possible.

A hot dry rock system at Fenton Hill, New Mexico, has tested the basic concept and proved it can technically work. Plenty of engineering work remains to be completed, though, and the economic feasibility is far from clear. High temperature gradients below Roosevelt Springs, Utah, and Clear Lake, California, make these locations excellent early candidates for exploiting this technology. Better deep-drilling techniques could open up the entire West as well as much of the Midwest and South for tapping into hot dry rock.

*Magma* represents an even more powerful and long-lasting geothermal energy source than hot dry rock. But it also poses serious technical and engineering challenges. In most parts of the United States, molten magma lies more than 20 miles below sea level. In areas of recent earthquake activity, however, huge reservoirs of magma may be trapped within range of modern drilling equipment. The concept sounds like the stuff of science fiction, but researchers in 1982 successfully drilled into a crusted-over magma lake in Hawaii. In this experiment, relatively cool fluids (twice the boiling point of water) circulating around the drill bit and pipes solidified a thin tube of the 1,000 degrees C magma, protecting the equipment from the intense heat and forming a stable drill hole.

In theory, 0.5 cubic mile of magma could run a 1,000 megawatt power plant for thirty years. However, transforming this concept from theory into practice will undoubtedly take several decades at minimum.

*Geopressured brines* discovered along the Gulf Coast in Texas, Louisiana, and Mississippi are an interesting blend of geothermal and fossil energy. The brines contain hot, salty water under high pressure. They also contain dissolved methane and other hydrocarbon gases, all

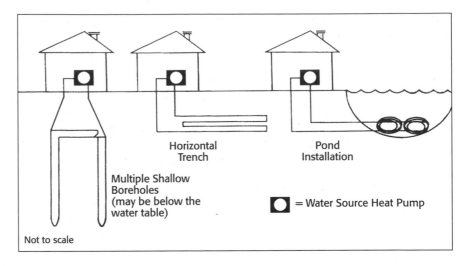

*Ground-coupled heat pump units are linked to the ground, or a pond or lake, by a heat exchanger, called a ground loop. The GCHP circulates a heat transfer fluid through the ground loop, absorbing the earth's natural warmth in winter and rejecting heat to the earth in summer. Ground-coupled systems are suitable for a wide range of climates and locations.*

Horizontal Trench

Pond Installation

Multiple Shallow Boreholes (may be below the water table)

= Water Source Heat Pump

Not to scale

that remains of ancient, decomposed marine life. Several test wells brought up brines that exceeded the boiling point of water (ranging from 130 degrees C to 260 degrees C) and that contained up to 100 cubic feet of gas per barrel. An experimental power plant at the Pleasant Bayou Well south of Houston generated electricity by burning the recovered methane and using the hot brine to vaporize a working fluid that was then funneled through a turbine. Though an industrial consortium has formed to develop this resource, its contribution to the U.S. energy supply in the near future will probably be small.

Unlike conventional furnaces, *ground source heat pumps* do not generate heat but merely *move* it from one place to another. This allows them to operate far more efficiently than traditional heating systems, using as little as 50 percent of the energy required by a natural gas or oil burner to heat a home. A recent study by the U.S. Environmental Protection Agency concluded that heat pumps represent the most efficient, least expensive to operate, and least polluting heating and cooling system on the market today.

Several meters below the ground surface, temperatures remain relatively constant throughout the year, thus providing ideal conditions for operation of a heat pump. During the winter, a heat pump works like a refrigerator in reverse: buried tubes carry a heat transfer fluid that attains the ground's temperature; it is pumped into a heat exchanger in the house and compressed, causing its temperature to rise dramatically. This heat is used in the house, causing the transfer fluid to contract and cool down. It is then pumped back outside to pick up more heat from the ground. During the summer, the process is reversed, and the ground acts as a sink for heat removed from air inside the building; just as the air acts as a sink for heat removed from the inside of a refrigerator.

**RESOURCE**     Current estimates of the United States' geothermal resources are merely educated guesses, but even the most conservative estimates suggest that we should be seriously developing geothermal energy. The United States Geological Survey places the amount of hydrothermal energy recoverable with current technology at the equivalent of 87 billion to 140 billion tons of coal. The total amount of potential energy in the ground may be twenty times larger. Geopressured brines alone could contain twice the energy in all our proven and unproven coal reserves. Hot dry rock and magma stores are even larger. A National Academy of Sciences review estimates that the U.S. geothermal resource may be thousands of times larger than its coal reserves.

**ADVANTAGES AND DISADVANTAGES**     In theory, geothermal energy can be as clean as solar or wind power, producing no air or water pollution. Any wastes produced, such as salts and minerals, can be pumped deep into the ground, where they will cause no harm. As with all renewable energy sources, geothermal is a home-grown resource that helps keep dollars in the region where it is produced and consumed.

Geothermal resources can be overexploited, however, meaning that heat is removed faster than natural reactions replace it. Developers of The Geysers, a major geothermal electricity-generating plant in California, have learned this lesson the hard way, watching output fall as the number of wells tapping the heat reservoir has grown. Water and steam released from the bowels of the earth usually contain dissolved minerals, salts, and gases that can become environmental hazards if not disposed of properly. In rare instances, removing large volumes of water from a geothermal well can make the ground above settle. Water or other fluids must be reinjected into the well to prevent this problem.

Finally, while geothermal resources are scattered around the country, some of the finest sit below wild and scenic areas. The Geysers plant abuts California's picturesque and productive wine country. Gargantuan geothermal fields appear to exist beneath Yellowstone National Park (they keep Old Faithful erupting!), Crater Lake in Oregon, and Volcanoes National Park in Hawaii. These resources will almost certainly not be used, making the development of other resources, such as hot dry rock, all the more important.

## Energy Storage

Some critics charge that renewable energy sources like the sun and wind have a fragile Achilles heel—they can't produce power all day long or increase their output on demand. While that is technically true, it's not a fatal drawback that should either constrain or impede the development of these energy resources. One solution is mixing a variety of renewable resources in the same region. For example, biomass-burning generators could take over for solar thermal generators when the sun goes down. In some parts of the country, night winds may perfectly complement photovoltaic systems. Adding a host of energy storage systems could further reduce, if not eliminate, problems of demand and reliability.

There's little mystery to energy storage; we already do it in many ways and for many tasks. We're all familiar with batteries, and most homes have storage tanks for hot water. A growing number of commercial buildings, like Chicago's massive Merchandise Mart, store "cool" by making ice at night when electricity is relatively cheap and using it for daytime air-conditioning. Several utilities pump water uphill into storage reservoirs at night and then let it run downhill through turbines during the daytime, increasing their electricity generating capacity for peak demand without relying on backup oil-burning plants.

Hydrogen represents a form of storage that futurists have been dreaming about for years. In the fabled hydrogen economy, the energy currency is a common, colorless gas electrically made from water that, when burned, produces water as a waste product.

**HOW IT WORKS** Energy can be stored in several ways—as chemical energy, potential energy, heat, or even as electricity. All are currently being used or investigated.

*Batteries* turn electricity into chemical energy and vice versa. The most common ones use easily reversed reactions between lead and sulfuric acid; more advanced batteries being developed for electric vehicles use nickel and iron, or molten sodium and sulfur. For many owners of small photovoltaic or wind systems, batteries turn a not-always-there resource into one that's constant and reliable. When the sun shines or the wind blows, electricity is used to charge the battery. After sunset or when the breezes stop, the battery automatically supplies electricity for running a refrigerator, lights, or a television set. Properly sizing battery storage is an important task for anyone depending on a wind turbine or photovoltaic array for electricity.

Passive solar design depends largely upon *thermal storage,* simply and efficiently storing heat in masonry walls or floors, an indoor pool, or a rock bed. These "heat sinks" gradually warm up, and stay warm for hours after the sun has set. Without such storage, the temperature in a passively heated home could skyrocket during the day and plummet at night. The concept of thermal storage isn't new—adobe made from sun-dried earth and straw has been used in construction for hundreds of years because it stores heat so well. The amount of heat that a kilogram of rock or water can store (its specific heat) is relatively low, meaning large masses are needed to store a lot. Researchers are testing a variety of substances that liquify or vaporize at low temperatures, since this process stores far more heat per unit weight than simple heating.* Leading candidates include paraffin and several salts. The trick is encapsulating them in building materials like roofing shingles or drywall so they don't dribble away when melted.

Thermal storage may also be feasible on a large scale. At Solar Two (see page 30), the heat generated by hundreds of mirrors will be stored in molten sodium, which can be used either immediately or hours later for generating electricity.

The ultimate in thermal storage sounds so simple, yet implausible. Why not store away summer's heat for winter heating? Several communities in Sweden are doing just that. Solar collectors mounted on buildings circulate sun-heated fluid through huge piles of buried and insulated crushed rock or underground ponds. Throughout the winter, this stored energy is used to supply a district heating system. A U.S. test of such seasonal storage is underway at the University of Massachusetts, Amherst, where a field of solar collectors will warm up waterlogged clay deposits. The stored heat will be used in a nearby gymnasium and arena. For a smaller-scale example, see EarthConnection, page 134.

Two utility-scale technologies, *compressed air energy storage* and *superconducting magnetic energy storage,* offer the possibility of storing huge quantities of renewably generated electricity and thus smoothing out daily variations in electricity production. The nation's first compressed air energy storage plant, completed in 1991, uses cheap

---

*Heating a kilogram of water 1 degree Celsius takes 1,000 calories of heat. Melting a kilogram of ice without raising its temperature, by comparison, requires 80,000 calories of heat.

night-time electricity to pump air into an airtight salt cavern 2,500 feet below Birmingham, Alabama. When the local utility needs extra electricity it basically pulls the cork and lets this compressed air blast out through a turbine.

Superconducting magnetic energy storage (SMES) is the only system that stores electricity as electricity without first converting it to chemical or potential energy. It involves pumping electricity into a huge coil of wire as large as half a mile across that has been cooled to below –270 degrees C. A small SMES system built for the city of Tacoma, Washington, operated successfully for a year. The federal departments of Energy and Defense have been working on a large, 10-megawatt system, and a Wisconsin company makes small SMES systems not much bigger than a home water heater that store enough electricity to smooth out brief power surges and safely shut down a factory's computers in case of a power failure.

*Pumped hydropower* is another method for storing massive amounts of energy. It operates on the same principle as natural hydropower, except the moving force is electricity rather than the sun. During off-peak hours, pumps push water into an elevated reservoir; it falls back down through turbines when electricity is needed. While this shuffle consumes more electricity than it produces, it allows a utility to use excess generating capacity at night and avoid bringing old oil- or coal-fired generating plants on-line when daytime demand for electricity peaks.

For storing renewable energy, *hydrogen* is hard to beat. This odorless, colorless gas is made by electrically splitting water into its two components, hydrogen and oxygen. Recombining (burning) the two releases heat and a single, nontoxic waste product—water. While it makes more sense to use the renewably generated electricity directly, sometimes that isn't possible. Hydrogen could be shipped long distances through existing natural-gas pipelines that have been slightly modified. This network already connects the Southwest, a prime area for photovoltaic or solar thermal electricity generation, with the Northeast. Hydrogen could also be used as a transportation fuel in modified internal combustion engines or fuel cells—battery-like devices that generate electricity when fed a constant stream of hydrogen.

# Impacts and Hidden Costs of Fossil Fuels

F ossil fuels—coal, oil, and natural gas—are America's primary source of energy, accounting for 85 percent of current U.S. fuel use. And Americans pay for their heavy use of these fuels with rapidly diminishing energy supplies for a rapidly expanding number of people, foreign wars to protect fuel supplies, dirtier and grimier cities, ever-worsening air quality, dead lakes and fouled waters, and increased incidence of heart disease, bronchitis, allergies, and other respiratory ailments. To adequately address this range of environmental and human health problems, the United States must switch from polluting, depletable fossil fuels to clean, renewable energy sources, such as sunlight and wind, that are regenerated at the same rate they are used.

In great part, this country's heavy reliance on fossil fuels to meet our electrical and heating needs continues because policymakers and consumers fail to take into account the deadly hidden costs of fossil fuel use. Some of the costs of using fossil fuels are obvious, such as labor to mine for coal or drill for oil, and these costs are included in monthly electric bills or in the purchase price of gasoline for cars. But other energy costs are not included in consumers' utility or gas bills, nor are they paid for by the companies that produce or sell the energy. These costs include human health problems caused by air pollution from the burning of coal and oil; damage to land from coal mining and to miners from black lung disease; environmental degradation caused by global warming, acid rain, and water pollution; national security costs of protecting foreign sources of oil; and a negative impact on the balance of trade and the U.S. deficit.

Since such costs are indirect and difficult to determine, they have traditionally remained external to the energy pricing system, and are thus often referred to as "externalities." But many studies have tried to quantify these costs. The American Lung Association, for example, claims that black lung disease is responsible for approximately four thousand deaths in the U.S. each year. A review study conducted by the Pace University Center for Environmental Legal Studies found that every ton of sulfur dioxide emitted into the atmosphere (mainly from coal-fired power plants) costs society an estimated $3,500 in health-related damage alone—and in 1991 the United States emitted 20.7 million tons of that pollutant. The Congressional Research Service estimates that the military cost of securing peacetime oil transportation routes is somewhere between $1 billion and $70 billion per year. Since the producers and the specific users of energy do not pay for these costs, however, society as a whole must pay for them. This pricing system masks the true costs of fossil fuels and results in serious damage to human health, the environment, and the economy.

## Environmental and Health Impacts of Fossil Fuel Use

Many of the environmental problems our country faces today result from our fossil fuel dependence. These impacts include global warming, air-quality deterioration, oil spills, and acid rain.

**GLOBAL WARMING**    Scientists have known for a century that certain gases in the atmosphere trap heat and keep the earth warm, much as the glass of a greenhouse keeps the air inside warm. This atmospheric blanket is in fact essential to life: without it, the earth would be so cold that it would be uninhabitable. However, industrial civilization is now adding to the concentrations of the "greenhouse gases" in the atmosphere. Although atmospheric scientists are divided over whether this increase is already causing a warming of the earth, they believe almost unanimously that such a warming will occur in the future.

The burning of fossil fuels for energy accounts for 60 percent of all greenhouse-gas emissions. Among such gases, one of the most important is carbon dioxide ($CO_2$); the United States alone spews out over five billion tons of carbon dioxide annually. Over the past 150 years, fossil fuel burning has resulted in more than a 25 percent increase in the amount of $CO_2$ in our atmosphere. Fossil fuels are also implicated in increased levels of atmospheric methane ($CH_4$) and nitrous oxide ($N_2O$), although they are not the major source of these gases.

A variety of sophisticated computer models predict that if atmospheric $CO_2$ levels continue to increase, our planet will become warmer in the next century. Projected temperature increases—ranging

from 3 degrees F to 5 degrees F—will result in a variety of environmental impacts, some more certain than others. One almost certain change is that the oceans will rise because polar ice sheets and alpine glaciers will partially melt and because warmer water takes up more space. Studies suggest that the sea-level will rise between 30 centimeters and 1 meter (1 to 3 feet) by the mid-twenty-first century, enough to hasten shore erosion, destroy irreplaceable wetlands, and contaminate water supplies and drainage systems with seawater. If the oceans continue to rise, many more low-lying coastal areas will be flooded.

*Several harmful pollutants are produced by fossil fuel combustion, contributing to global warming and possible climate change, ever-worsening air quality, dead lakes and fouled waters, and serious health problems. These are some of the "hidden costs" of burning coal, oil, and gas.*
Photo credit: Herb Rich.

Major changes in weather patterns are also expected. Overall, average precipitation around the world will rise—but not necessarily where and when it is needed most. In the interiors of continents, the weather may actually become drier in the summer, causing more frequent droughts. As the oceans warm, the severity and frequency of tropical storms and hurricanes could increase. Changes such as these could have a serious impact on agriculture and plant and animal life, as well as on human populations. Looming even more grimly behind these identifiable dangers is the threat of massive and unpredictable changes in climate.

Although it is impossible to predict exactly what will and will not happen, virtually all scientists agree that this buildup of greenhouse gases has the potential to produce dramatic climate changes. Because it will take decades to reverse the greenhouse-gas buildup in the atmosphere, many eminent scientists are calling upon the world's political leaders to take immediate steps to reduce greenhouse-gas emissions, while simultaneously calling for more research on global warming.

Above all, world consumption of fossil fuels must be reduced. The United States—which has just 5 percent of the world's population yet consumes nearly a quarter of the world's energy—must take the lead in cutting fossil fuel use.

**AIR POLLUTION**    Several important pollutants are produced by fossil fuel combustion: carbon monoxide (CO), nitrogen oxides ($NO_x$), sulfur oxides ($SO_x$), and hydrocarbons (HC). Ash and other particulates contribute to air pollution, and $NO_x$ and HC can combine in the lower atmosphere to form tropospheric ozone ($O_3$), the major constituent of smog.

Carbon monoxide is a gas formed as a by-product during the incomplete combustion of fossil fuels. Cars and trucks are the primary source of CO emissions. Exposure to CO can cause headaches and place additional stress on people with heart disease. Two oxides of nitrogen—nitrogen dioxide, ($NO_2$) and nitric oxide (NO)—are also formed in combustion. Nitrogen oxides appear as yellowish-brown clouds over many city skylines, and can irritate the lungs, cause bronchitis and pneumonia, and decrease resistance to respiratory infections. They also lead to the formation of ozone, which is toxic in the lower atmosphere. (In the stratosphere, ozone sheilds us from the sun's harmful rays.) The transportation sector is responsible for close to half of the $NO_x$ emissions in the United States; power plants produce most of the rest.

Sulfur oxides are produced by the oxidization of sulfur in fuel. Utilities that use coal to generate electricity produce two-thirds of the nation's sulfur dioxide ($SO_2$) emissions. Nitrogen oxides and sulfur oxides are important constituents of acid rain. These gases combine with water vapor in clouds to form sulfuric and nitric acids, which precipitate as acid rain and snow. As the acids accumulate, lakes and rivers become too acidic for plant and animal life. Acid rain also dissolves building materials such as limestone and mortar, and leaches nutrients out of the soil so crops are damaged.

Hydrocarbons (HC) are a broad class of pollutants made up of hundreds of specific compounds containing carbon and hydrogen. The simplest hydrocarbon, methane ($CH_4$), does not readily react with $NO_2$ to form smog, but all other hydrocarbons do. Hydrocarbons are emitted from human-made sources such as auto and truck exhaust, evaporation of gasoline and solvents, and petroleum refining.

The white haze that can be seen over many cities is tropospheric ozone, more commonly known as smog. This gas is not emitted directly into the air; rather, it is formed when ozone precursors—mainly nonmethane hydrocarbons and nitrogen oxides—react in the presence of heat and sunlight. Human exposure to ozone can produce shortness of breath and, over time, permanent lung damage. Research shows that ozone may be harmful at levels even lower than the current federal air standard. In addition, ozone reduces crop yields by as much as 5 percent in parts of the United States.

Finally, fossil fuel use also produces particulates, including dust, soot, smoke, and other suspended matter, which are respiratory irritants. Particulates may contribute to acid rain formation.

*Cleaning up Prince William Sound after the Exxon Valdez oil spill.*
Photo credit: Oil Information Center, Alaska.

**WATER AND LAND POLLUTION**    One source of water pollution is the production, transportation, and use of oil. Oil spills and leaks from ships and pipelines foul waterways and their surrounding shores, killing thousands of fish and the animals that feed on them or live in the water. Coal mining also contributes to water pollution. Coal contains pyrite, a sulfur compound; as water washes through mines, this compound forms a dilute acid, which is then washed into nearby rivers and streams.

Coal mining, especially strip mining, drastically affects the area that is being mined. Characteristically, the material closest to the coal is acidic and often sulfur bearing; these materials can change the pH balance of the ecosystem and disrupt flora and fauna. After the mining is completed, the land will remain barren unless special precautions are taken to ensure that proper topsoil is used when the area is replanted. Materials other than coal like sulfur-bearing compounds and heavy metals can be brought to the surface in the coal mining process, and these are left behind as solid wastes. As the coal itself is washed, more metal and acid-forming chemical waste material is left. Finally, as the coal is burned, the remaining ash is left as a waste product.

**THERMAL POLLUTION**    During the electricity generation process, burning fossil fuels produce heat energy, some of which is used to generate electricity. Because the process is inefficient—as much as 70 percent of the energy in the fuel is lost—much of the heat is released to the atmosphere or to water that is used as a coolant. Heated air is not a problem, but heated water, once returned to rivers or lakes, can upset the aquatic ecosystem.

## National Security Impacts of Fossil Fuel Use

Our nation's fossil fuel dependence means that, to ensure our supply, we may be forced to protect foreign sources of oil. The Persian Gulf War is a perfect example: in 1991 U.S. troops were sent to the Persian Gulf, in large part to guard against a possible cutoff of our oil supply. Not only were billions of dollars spent in protecting the oil, but thousands of lives were lost on both sides as well. Although the war is over, through taxes we are continuing to pay for protecting oil supplies with our armed forces. As domestic fossil fuel supplies dwindle, the potential for future "energy wars" increases.

## Economic Impacts of Fossil Fuel Use

Reliance on oil from the Middle East creates a danger of fuel price shocks or shortages if the supply is disrupted. Today, about one-third of our oil comes from the Middle East; by 2030, if we do not change our energy policy, we may be relying on Middle East oil for two-thirds of our supply. Oil imports now account for about 40 percent of the U.S. trade deficit.

In addition to the national picture, a state-by-state examination reveals that the vast majority of states are net energy importers, sending millions of energy and tax dollars out of state—and jobs along with dollars. Even some of the traditionally energy-rich states, such as Texas, now find themselves unable to meet their own needs.

And looking at the economics of energy from an international perspective, U.S. energy use is once again notably disadvantageous. Japan and Germany (at least the former West Germany) are twice as energy-efficient as the United States, which is both good for the planet and good for the economies of these two countries, since the lower cost of manufacturing makes Japanese and German products more attractive in the world market. Moreover, the U.S. runs the risk of losing out on future economic opportunities because we are failing to invest in cutting-edge energy technology development. All in all, our country is suffering great economic loss from our present energy policies and practices—yet another example of the hidden costs of dependence on fossil fuels.

# *Resource Guide*

This guide is designed to direct you to the information and services you need to harness renewable energy in your community. Whether or not you have a specific project in mind, you may want to start by researching the relevant federal, state, and local regulations and programs affecting energy in your area. There may be special incentives or programs to promote renewable energy or energy efficiency which could help your project. Also important are the policies and plans of utilities serving your area and their programs for incorporating renewables. Many utilities are using renewables and efficiency as a demand-side management tool, or as a way of providing energy to remote customers. Finding out more about your community's situation will give you an idea of what issues you might address and the directions your project could take.

As you investigate potentially useful organizations, you might also look for individual people who will be allies in your quest for renewable energy. It is individual people who really make things happen. While this Resource Guide lists the most important players on the national level, it also describes the groups which will likely exist in your local area. At the end of the guide is a list of publications which may be helpful to you.

## A. NATIONAL RESOURCES

### 1. Federal Government

Federal government agencies may be able to provide:
• Information, technical assistance, and further resources
• Information on regulations, and how to follow them
• Funding and other incentive programs

**Department of Energy (DOE)**
1000 Independence Avenue SW
Washington, DC 20585
202-586-5000
DOE coordinates and administers the energy functions of the U.S. government, including: long-term, high-risk research and development of energy technologies; marketing federal power; energy conservation; the nuclear weapons program; energy regulatory programs; and a central energy data collection and analysis program. DOE has ten regional offices around the country, and a number of specialized organizations within it, the most relevant of which are:

**Conservation and Renewable Energy Inquiry and Referral Service (CAREIRS)**
(See Energy Efficiency and Renewable Energy Clearinghouse)

**Design Assistance Center**
**Sandia National Lab**
Albuquerque, New Mexico 87185
505-844-8066
The Design Assistance Center was established to accelerate the adoption of photovoltaic technology. It serves as a non-commercial source of expertise on

photovoltaics for utilities, public utility commissions, and industry.

### Energy Efficiency and Renewable Energy Clearinghouse (EREC)
PO Box 3048
Merrifield, Virginia 22116
EREC provides information, technical assistance, and referrals on energy efficiency and renewable energy technologies. The staff can perform on-line searches of an extensive information database in order to give callers information over the phone, and/or to prepare a written response and package of relevant publications. As of January 1994, EREC replaces two previous services: The Conservation and Renewable Energy Inquiry and Referral Service (CAREIRS) and the National Appropriate Technology Assistance Service (NATAS).

### Energy Information Administration (EIA)
### National Energy Information Center (NEIC)
U.S. Department of Energy (DOE) EI-231
Forrestal Building, Room 1F 048
Washington, DC 20585
202-586-8800
EIA maintains a comprehensive database of energy information relevant to energy resources and reserves, production, demand, and technologies; and related financial and statistical information relevant to the adequacy of energy resources to meet the nation's demand in the near and long-term future. NEIC is the public-relations arm of EIA. It catalogs and distributes energy data, acts as a clearinghouse for statistical information on energy, and makes referrals for technical information. It provides one-page information sheets on renewable energy sources.

### Federal Energy Regulatory Commission (FERC)
825 N. Capitol Street NE
Washington, DC 20426
202-357-9082
FERC regulates the interstate aspects of the electric and natural-gas industries and establishes rates for transporting oil by pipeline. Issues and enforces licenses for the construction and operation of non-federal hydroelectric power projects. FERC also advises federal agencies on the merits of proposed water development projects.

### National Appropriate Technology Assistance Service (NATAS)
(See Energy Efficiency and Renewable Energy Clearinghouse)

### National Renewable Energy Laboratory (NREL)
### (formerly Solar Energy Research Institute)
1617 Cole Boulevard
Golden, Colorado 80401
303-231-1000
NREL is dedicated to supporting those applications of solar energy that are commercially attractive and environmentally acceptable. The staff conducts research in various phases of basic solar energy technologies and produces numerous publications.

### Oak Ridge National Laboratory
PO Box 2008
Oak Ridge, Tennessee 37831-6266
615-574-4160 fax: 615-574-0595
Among the Oak Ridge National Laboratory's research programs has been extensive work on biomass and on utility-integrated resource planning.

### Solar Energy Research Institute (SERI)
(See National Renewable Energy Laboratory (NREL))

### Tennessee Valley Authority (TVA)
4S 124X Missionary Ridge Place
Chattanooga, Tennessee 37402-2801
615-751-7521
The TVA is a federal corporation responsible for developing the natural resources of the Tennessee Valley and adjoining regions. Its primary mission is power production, which it does with a combination of renewable, fossil, and nuclear sources. TVA has been involved in efficiency and renewable energy programs since the mid–1970s.

### U.S. Congress
Members of Congress may have aides who focus on

energy issues or may serve on a committee which deals with energy issues.

## 2. National Trade Associations

Trade associations have publications and may do lobbying and develop demonstration projects.

**American Biofuels Association (ABA)**
1925 N. Lynn Street, Suite 1000
Arlington, Virginia 22209
703-522-3392 fax: 703-522-4193
ABA promotes biofuels within existing and developing industries. It provides information to federal and state bodies, and the environmental, public interest, and farm communities. ABA publishes a monthly newsletter, *Biodiesel Alert.*

**American Hydrogen Association (AHA)**
PO Box 15075
Phoenix, Arizona 85060-5075
602-438-8005
Formed to support the International Association for Hydrogen Energy (IAHE), AHA holds workshops and educational briefings for community, industrial, and environmental groups. AHA translates technical reports from IAHE's *Journal* into a form accessible to lay people.

**American Institute of Architects (AIA)**
1735 New York Avenue NW
Washington, DC 20006
202-626-7300 fax: 202-626-7518
AIA is the professional organization for architects in the U.S. It has a Center for the Environment which serves as an architect's clearinghouse for environmental information, and produces various publications. AIA also has a Committee on the Environment, a members' forum for the discussion of environmentally sustainable architecture.

**American Public Power Association (APPA)**
2301 M Street NW
Washington, DC 20037-1484
202-467-2900 fax: 202-467-2910
APPA serves the nation's community-owned, non-profit electric utilities. APPA represents its members before Congress, federal agencies, and the courts; provides educational programs; has various publications; funds energy research and development projects; and acts as a resource for government, the media, public interest organizations, and the public.

**American Society of Heating, Refrigerating, and Air-Conditioning Engineers (ASHRAE)**
1791 Tullie Circle NE
Atlanta, Georgia 30329
404-636-8400 fax: 404-321-5478
ASHRAE works to advance the arts and sciences of heating, refrigeration, air-conditioning, and ventilation. It sponsors research, develops standards, publishes technical data, and organizes meetings and educational activities. ASHRAE may be able to provide technical assistance on matters regarding climate control in the built environment.

**American Wind Energy Association (AWEA)**
777 N. Capitol Street NE, Suite 805
Washington, DC 20002
202-408-8988
AWEA represents the wind energy industry. They publish a newsletter and can supply information on wind power and turbine manufacturers.

**Business Council for a Sustainable Energy Future**
1725 K Street NW, Suite 509
Washington, DC 20006
202-785-0507 fax: 202-785-0514
The Business Council is comprised of business leaders from the energy efficiency, renewable energy, natural gas, and utility industries who work with senior officials from major environmental organizations. The Council supports energy-related policies and programs which will enable the United States to achieve a sustainable pattern of energy production and use that simultaneously contributes to the nation's economic, environmental, and national security goals.

**Earth Energy Association (EEA)**
1701 K Street NW, Suite 400
Washington, DC 20006
202-289-0868 fax: 202-223-1393
EEA serves as a liaison for manufacturers, utilities, engineering and contracting firms, government agencies, and other institutions to promote geothermal heating and cooling technology. EEA serves as a forum for the exchange of information among its members, and endeavors to increase government and general public awareness of the benefits to be derived from geothermal heat pump technologies.

**Edison Electric Institute (EEI)**
1111 19th Street NW
Washington, DC 20036-3691
202-778-6660
The EEI is the association of U.S. investor-owned electric utilities and industry affiliates. It provides reports and other materials on aspects of electric energy.

**Electric Power Research Institute (EPRI)**
PO Box 10412
Palo Alto, California 94303
415-855-2000 fax: 415-855-2266
EPRI is a research organization funded by its 700 member utilities which works on the generation, transmission, and use of electricity. EPRI produces numerous publications relating to these topics.

**Fiber Fuels Institute (FFI)**
Natural Resources Research Institute
University of Minnesota, Duluth
5031 Miller Trunk Highway
Duluth, Minnesota 55811
FFI is a trade association of companies and organizations involved in the use of wood, agricultural residues, paper, and other biomass for fuel. It promotes pellets and other fiber fuels in homes, businesses, institutions, and industrial plants.

**Geothermal Resources Association (GRA)**
1207 Potomac Street NW
Washington, DC 20007
202-333-5657 fax: 202-333-5001
GRA is comprised of fifteen of the nation's top companies making electricity from geothermal steam. It promotes the use of geothermal energy by publicizing the benefits to regulators, lawmakers, and the administration. Their quarterly newsletter is called *Geothermal News*.

**International District Heating and Cooling Association (IDHCA)**
1101 Connecticut Avenue NW, Suite 700
Washington, DC 20036
202-429-5111
IDHCA represents entities engaged in supplying thermal energy in the form of steam, and hot and chilled water for heating, cooling, and process use in community energy systems. IDHCA's membership consists of utilities, municipalities, universities, hospitals, military bases, airports, industrial parks, and other physical plant systems which use district heating, as well as architects, engineers, and manufacturers of equipment for district heating systems.

**National Association of Energy Service Companies (NAESCO)**
1200 G Street NW, Suite 760
Washington, DC 20005
202-347-0419 fax: 202-393-0336
Energy service companies (ESCOs) get paid for the energy savings delivered to the customer, typically over a five- to fifteen-year period. NAESCO represents such companies and supports the implementation of market and pricing signals, and other incentives for utilities and ESCOs, which reflect the social and economic value of energy efficiency.

**National Association of Home Builders (NAHB)**
1201 15th Street NW
Washington, DC 20005
202-822-0200
NAHB represents the home building industry and has publications on sustainable housing and other relevant topics. Its Research Center investigates new technologies and methods for the home building industry.

**National Hydropower Association (NHA)**
555 13th Street NW
Washington, DC
202-637-8115
NHA promotes hydroelectric power to the public, policy makers, and regulators through public outreach and advocacy. NHA produces materials regarding legislative, legal, and technological developments in hydropower.

**National Rural Electric Cooperative Association (NRECA)**
1800 Massachusetts Avenue, NW
Washington, DC 20036
202-857-9500
This national service organization produces publications and provides technical assistance and support to rural electric cooperatives.

**National Wood Energy Association (NWEA)**
777 N. Capitol Street NE
Washington, DC 20002
202-408-0664
NWEA is the trade association for the wood energy industry. It seeks to encourage the responsible development of biomass, and focus public and legislative attention on this resource. Its quarterly trade journal is called *Biologue*.

**Passive Solar Industries Council (PSIC)**
1511 K Street, NW, Suite 600
Washington, DC 20005
202-628-7400
PSIC is a national association devoted to providing practical information on energy-conscious passive solar design and construction. PSIC sponsors a national workshop series, and provides guidelines for residential new construction and remodeling, as well as small commercial materials available in print and computerized versions.

**Photovoltaics for Utilities (PV4U)**
15 Hayden Street
Boston, Massachusetts 02131-4013
617-323-7377 fax: 617-325-6738
PV4U is comprised of key players from the photovoltaic industry and from utility, regulatory, government, and consumer communities. It works to create markets for photovoltaics, functioning through working groups in a number of states.

**Renewable Fuels Association (RFA)**
1 Massachusetts Avenue NW, Suite 820
Washington, DC 20001
202-289-3835 fax: 202-289-7519
RFA is a trade association representing the domestic ethanol industry.

**Solar Energy Industries Association (SEIA)**
777 N. Capitol Street NE, Suite 805
Washington, DC 20002-4226
202-408-0660 fax: 202-408-8536
Representing the solar energy business community, SEIA provides support for the industry and information to the public. The Association's work includes technical and economic research, government affairs, publications, conferences, and workshops. SEIA publishes the *Solar Industry Journal*, and a catalog called *Renewable Energy Publications*.

**United States Export Council for Renewable Energy (US/ECRE)**
122 C Street, NW, Suite 400
Washington, DC 20001
202-383-2550 fax: 202-383-2555
A consortium of the six U.S. trade associations that promote the use of alcohol fuels, energy efficiency, hydropower, passive solar, photovoltaics, solar thermal, wind, and wood. They can help identify the companies best suited to manufacture, distribute, and maintain renewable energy systems through direct sales, joint ventures, and technology transfer agreements. Through workshops and conferences, the Council trains practitioners in various aspects of renewable energy.

**Utility Photo Voltaic Group (UPVG)**
1800 M Street NW, Suite 300
Washington, DC 20036
202-857-0898 fax: 202-223-5537
A group of seventy-six electric utilities, serving 40 percent of U.S. customers, working to expand the

practical application of photovoltaic electricity. UPVG works to increase the use of both small-scale and larger applications of PV.

**Utility Wind Interest Group (UWIG)**
Electric Power Research Institute
3412 Hillview Avenue
Palo Alto, California 94304
415-855-2159
In conjunction with the Electric Power Research Institute and the U.S. Department of Energy, the utilities of UWIG support the appropriate integration of wind technology for utility applications.

## 3. National Public Interest Groups

This section contains groups which represent the public, or some segment of it, rather than the government or businesses. We have listed organizations which work on renewable energy and energy efficiency, but many of these groups also work on other issues. Most engage in one or more of the following activities:

- Education and outreach to inform the public about important issues
- Advocacy, lobbying, and legal action directed towards government and other power-bases to change formal and informal rules.
- Research to guide the above activities.

**Alliance to Save Energy (ASE)**
1725 K Street NW, Suite 914
Washington, DC 20006
202-857-0666 fax: 202-331-9588
This nonprofit coalition of government, business, environmental, consumer, and labor leaders is dedicated to increasing the efficiency of energy use. The alliance conducts research, organizes pilot projects, develops educational programs, and formulates policy initiatives.

**American Council for an Energy-Efficient Economy (ACEEE)**
1001 Connecticut Avenue NW
Washington, DC 20036
202-429-8873 fax: 202-429-2248

ACEEE is dedicated to advancing energy-conserving technologies and policies. They advise governments and utilities on techniques for improving energy efficiency; research and prepare in-depth studies of energy-efficient technologies, policies, and related issues; organize conferences for researchers, practitioners, and policy-makers; and publish and distribute books, conference proceedings, and reports. They also publish a very useful guide to energy-efficient appliances.

**American Solar Energy Society (ASES)**
2400 Central Avenue B-1
Boulder, Colorado 80301
303-443-3130 fax: 303-443-3212
ASES is the national society for professionals and others involved in the field of solar energy. It provides forums for exchange of information on solar energy applications and research, publishes general and technical information on solar technologies, and promotes education in fields related to solar energy. The ASES magazine is *Solar Today*.

**Biomass Energy Research Association (BERA)**
1825 K Street NW, Suite 503
Washington, DC 20006
202-382-5595; 708-785-2856 answering service
1-800-247-1755 fax: 202-233-4625
A nonprofit membership organization working to develop and commercialize biomass energy systems. BERA supports public and private research, facilitates technology transfer, runs a speakers bureau, and has many publications. They have a directory of U.S. renewable energy hardware and systems.

**Campaign for New Transportation Priorities (CNTP)**
900 2nd Street NE, Suite 308
Washington, DC 20002
202-408-8362
A coalition of fifty organizations committed to bringing about a more balanced transportation policy through greater funding of intercity rail, urban mass transit, and other energy-efficient alternatives to driving.

Center for Environment, Commerce, and Energy
(CE)
African American Environmentalist Association
(AAEA)
317 Pennsylvania Avenue SE
Washington, DC 20003
202-543-3939
CE is an organization dedicated to protecting the environment, enhancing human ecology, promoting the efficient use of natural resources, and increasing African American participation in the environmental movement. AAEA is its membership arm, publishing *The African American Environmentalist,* and providing a database of the latest environmental issues facing the African American community.


Center for Global Change (CGC)
University of Maryland
Executive Building, Suite 401
7100 Baltimore Avenue
College Park, Maryland 20740
301-403-4165
The Center for Global Change maintains a database on state and local legislation and regulations relating to global climate change, including renewables and energy efficiency. It also conducts research on environmental policies, sponsors conferences, and produces several publications.


Citizen Action
1300 Connecticut Avenue NW
Washington, DC 20036
202-857-5153
Citizen Action lobbies for legislation about energy, the environment, and health care. It is sponsoring a state-centered, utility reform campaign in twelve states that pushes for the increased use of renewables and energy efficiency. Citizen Action produces various publications, including a report of the current energy policies of each state, and recommendations for future policies.

Electric Auto Association (EAA)
1249 Lane Street
Belmont, California 94002
415-591-6698
EAA's engineers, technicians, and hobbyists promote the development and use of electric vehicles. There are sixteen state and six regional chapters.


Geothermal Resources Council (GRC)
PO Box 1350
Davis, California 95617
916-758-2360
A membership organization which promotes geothermal energy, provides information, and encourages research, exploration, development, and understanding of the resource. The Council organizes workshops and seminars, has a wide range of publications, and publishes the *Bulletin* monthly.


Institute for Local Self-Reliance (ILSR)
2425 18th Street, NW
Washington, DC 20009
202-232-4108
The Institute provides research and technical assistance to local governments and citizens' organizations in the areas of recycling, energy, economic development, and agriculture.


National Audubon Society (NAS)
950 3rd Avenue
New York, New York 10022
212-832-3200 fax: 212-593-6254
Among this national environmental organization's many programs is the National Audubon Solar Brigade, an outreach program to encourage people to demand that their utilities increase the use of solar energy.


National Center for Appropriate Technology
(NCAT)
3040 Continental Drive
PO Box 3838
Butte, Montana 59702
406-494-4572
NCAT is a national organization dedicated to the mission of applying appropriate technologies to as-

sist low-income people. NCAT promotes energy efficiency and sustainable agriculture, proposes programs for appropriate technology to Congress, and publishes a catalog of information products on appropriate technology. In conjunction with the U.S. Department of Energy, NCAT runs the National Appropriate Technology Assistance Service.

## Public Citizen
215 Pennsylvania Avenue SE
Washington, DC 20003
202-546-4996 fax: 202-547-7392
Public Citizen has fought for consumer rights in the marketplace, for safe products, for a healthy environment, for clean and safe energy sources, and for corporate and government accountability. The Critical Mass Energy Project is the energy policy arm of the organization.

## Renew America
1400 16th Street NW, Suite 710
Washington, DC 20036-2217
202-232-2252
A membership organization promoting public involvement in the creation of a sustainable future. Their Searching for Success program awards and keeps track of outstanding environmental programs across the country, and their State of the States report is an annual evaluation and comparison of environmental programs and policies in each state.

## Rocky Mountain Institute (RMI)
1739 Snowmass Creek Road
Old Snowmass, Colorado 81654
303-927-3851 fax: 303-927-4178
RMI researches and informs the public about efficient, sustainable resource use, and has departments of energy, water, economic renewal, agriculture, and security. Its analyses focus on new technologies to reduce energy needs, and it supplies this information to policy-makers, builders, designers, and utility companies.

## Safe Energy Communications Council (SECC)
1717 Massachusetts Avenue NW
Washington, DC 20036
202-483-8491
The SECC is a coalition of national energy, environmental, and public interest media groups. It works to increase public awareness about the ability of our energy efficiency and renewable energy sources to meet an increasing share of our nation's energy needs, and of the serious economic and environmental liabilities of nuclear power.

## Solar Box Cookers International (SBCI)
1724 11th Street
Sacramento, California 95814
916-444-6616 fax: 916-447-8689
SBCI provides educational materials and training programs about solar box cookers, and works with other organizations that are promoting their distribution.

## Sun Day Campaign
315 Circle Avenue #2
Tacoma Park, Maryland 20912-4836
301-270-2258 fax: 301-891-2866
Sun Day is a nonprofit network of over 600 citizen groups, businesses, government agencies, and other organizations working to promote renewable energy and energy efficient technologies. Each April, activities across the country promote these energy futures in SUN DAY events.

## Union of Concerned Scientists (UCS)
2 Brattle Square
Cambridge, Massachusetts 02238
617-547-5552 fax: 617-864-9405
UCS is a nonprofit group which has formed an alliance between leading scientists and committed citizens dedicated to advancing responsible public policies in areas where technology plays a critical role. Renewable energy is a major focus of UCS's research, publications, and advocacy.

## US Public Interest Research Group (USPIRG)
215 Pennsylvania Avenue SE
Washington, DC 20003
202-546-9707
An environmental and consumer advocacy organization that helps influence new legislation and works for government reform. It has offices in many states around the country which are primarily staffed by college students. Some of the state and local chapters address energy issues.

## World Resources Institute (WRI)
1709 New York Avenue NW, Suite 700
Washington, DC 20006
202-638-6300 fax: 202-638-0036
WRI is an independent research and policy institute that has produced a number of reports and publications on renewables and other energy issues.

# B. NATIONAL ORGANIZATIONS REPRESENTING LOCAL INTERESTS

## Interstate Solar Council (ISC)
900 American Center Building
Saint Paul, Minnesota 55101
612-296-4737
This coalition of renewable energy program managers from twenty-six state energy offices promotes least-cost utility planning, certifies solar equipment, and co-sponsors Solar Energy Industry Association annual conferences.

## National Association of Regulatory Utility Commissioners (NARUC)
PO Box 684
Washington, DC 20044
202-898-2200
NARUC is composed primarily of state public utility commissioners, and seeks to improve the quality and effectiveness of utility regulation. The Association also has membership drawn from federal, state, and local government bodies, and is divided into a number of committees; the Committee on Energy Conservation has a Subcommittee on Renewable Energy which focuses on the regulatory aspects of renew-

ables. The Association's weekly *Bulletin* keeps its members up to date and it has produced a technical guide, *Investing in the Future: A Regulator's Guide to Renewables.*

## National Association of State Energy Officials (NASEO)
1615 M Street NW, Suite 810
Washington, DC 20036
202-546-220 fax: 202-546-1799
NASEO's membership includes representatives from fifty-three of the fifty-six state and territorial energy offices. NASEO represents the states at the federal and congressional level and keeps its membership informed on developments in Washington, DC.

## National Association of State Utility Consumer Advocates (NASUCA)
1133 15th Street NW, Suite 575
Washington, DC 20005
202-727-3908
NASUCA is a national organization of utility ratepayer advocates representing consumers served by investor-owned gas, telephone, electric, and water companies. Member offices are independent of state regulatory commissions and are either separate organizations or part of larger departments, often the Attorney General's office. Various publications and a computer network keep members and other interested parties up to date. NASUCA has a subcommittee and a number of workshops dedicated to renewable energy.

## National Conference of State Legislatures (NCSL)
Denver Office
1560 Broadway, Suite 700
Denver, Colorado 80202
303-830-2200 fax: 303-863-8003

Washington Office
444 N. Capital Street NW, Suite 515
Washington, DC 20001
202-624-5400
NCSL is a nonprofit, bipartisan organization dedicated to serving the lawmakers and staff of the nation's fifty states, its commonwealths, and territories.

NCSL's energy work includes familiarizing state legislators with energy-efficiency and renewable energy options, and helping states develop programs for efficiency and renewable energy. NCSL is a clearinghouse for information on a range of energy and environmental issues and a source for technical assistance, research, and consulting services. It holds numerous meetings and seminars on these issues, and publishes a newsletter and various reports.

## C. REGIONAL, STATE, AND LOCAL RESOURCES

### 1. Regional, State, and Local Government

#### REGIONAL GOVERNMENT

Regional Biomass Programs can provide assistance and funding for biomass projects. They are funded by the U.S. Department of Energy and administered by state or regional government bodies, as listed below. Each regional office will be able to connect you to the state and technical consultants.

**Great Lakes Region**
Fred Kuzel
Council of Great Lakes Governors
35 East Wacker Drive #1850
Chicago, Illinois 60601
312-408-0177 fax: 312-408-0038

**Northeast Region**
Phil Lusk
CONEG Policy Research Center, Inc.
400 N. Capitol Street NW, Suite 382
Washington, DC 20001
202-624-8454 fax: 202-624-8463

**Northwest Region**
Pat Fox, Program Manager
Bonneville Power Administration
Box 3621, Portland, Oregon 97208
503-230-3449 fax: 503-230-4973

**Southeast Region**
Phillip Badger
Tennessee Valley Authority
Southeast Regional Biomass Energy Program
435 Chemical Engineering Building
Muscle Shoals, Alabama 35660
205-386-3086 fax: 205-386-2963

**Western Region**
Western Area Power Authority
1637 Cole Boulevard, PO Box 3402
Golden, Colorado 80401
303-231-1615 fax: 303-231-1632

#### STATE GOVERNMENT

**State legislature:** The members of the legislature's energy committees are likely to be the legislators most interested in energy, and they may have aides knowledgeable on energy issues.

**Governor's office:** Special programs on renewables or related issues may be run out of this office. For instance, in California the governor's office runs the California Energy Extension Service, which works to help small business, schools, and Native American tribes learn about and harness their renewable energy resources.

**State Energy Office:** A resource for information, technical assistance, and knowledge of other key groups in government, business, and public interest sectors. Programs to help fund renewable energy and efficiency projects, or develop related technologies, are often run by this office. For instance, when a group at Slippery Rock University in Pennsylvania created a "Master of Science in Sustainable Systems" program, they wanted to house it in a sustainable, renewably powered building. The Pennsylvania Energy Office provided $55,000 towards the realization of this goal (see page 150).

**Department of Environment/Ecology/Natural Resources or equivalent:** This department may have an interest in renewable energy as a means to protect the environment. For example, the Natural Resource and Environmental Protection Cabinet (NREPC) of Kentucky used the Kentucky Alternate Energy Development Fund to enable the Freeman Corporation in Winchester to make the capital investments necessary to convert its energy supply from natural gas to wood waste generated by its veneer manufacturing facility (see page 89).

**Economic Development/Commerce Department:** This department may be interested in promoting renewable energy industries or the use of renewables within the state. This department in Mississippi sponsored a comprehensive survey of the state's biomass resources, and then established a loan program to help businesses take advantage of the potential identified in the report (see page 80).

**Public Service Commission or Public Utilities Commission:** The commission regulates utilities, decides whether new generating capacity is needed, and interprets the Public Utilities Regulatory Policy Act (PURPA). One of the important aspects of PURPA is the requirement that independent power producers be allowed to sell power back to the grid. This means renewable energy sources don't necessarily have to be matched to a specific energy user, but can feed back into the grid, receiving payment from the utility. People on the commission will be knowledgeable about the energy situation in the state and may be personally interested in seeing renewable energy become mainstream. In the Beaver Island energy project (see page 151) the Public Service Commission was instrumental in linking the various actors in the project together.

**Special Organizations:** Your state may have established its own organizations to develop and promote renewable energy. For instance, in 1974, Florida created the Florida Solar Energy Center, now nationally recognized for comprehensive programs in alternative energy research and development.

Depending on your project, work with renewable energy may also bring you in contact with people in government organizations dealing with planning, public works, air quality, buildings and construction, natural resources, and solid waste.

**LOCAL GOVERNMENT**
Renewable energy advocates may need to know about and/or work with the local conservation commission, planning department, energy committee, or zoning board.

## 2. Regional, State, and Local Trade Associations and Businesses

If you are planning to acquire hardware for your project, the best starting points are probably the national trade associations for the energy types you are investigating. They will be able to provide you with relevant information and lead you to state and local businesses if they exist.

The National Appropriate Technology Assistance Service, your State Energy Office, and the Chamber of Commerce should be able to help you locate other relevant organizations. There may also be guidebooks about energy-related businesses in your state.

Much of the hardware needed in most renewable energy projects is general purpose and can be acquired from traditional sources (e.g. electrical, plumbing, or combustion equipment). But for planning, design, and construction, it may be necessary to find people familiar with renewable energy concepts. With a bit of research, you should be able to find architects, contractors, consultants, and suppliers who are versed in renewable energy applications.

Your local utilities are, of course, key players in the energy arena. They may have forward-thinking people in important positions who can be instrumental in advancing renewable energy. If your utility is one of these, start networking. If not, there may be little help and even some obstruction from this quarter. In the latter case, by educating people in your utility, you will advance renewable energy in the long term.

# 3. Regional, State, and Local Public Interest Groups

## REGIONAL GROUPS

**Coalition for Energy Efficiency and Renewable Technologies (CEERT)**
V. John White Associates
Research and Education Fund
1100 11th Street, Suite 321
Sacramento, California 95814
916-447-7983
CEERT is a Sacramento-based coalition of major renewable energy companies, environmental organizations, and public interest groups. It carries out policy research, regulatory and legislative advocacy, and public education in support of energy efficiency and renewable energy.

**Great Lakes Renewable Energy Association (GLREA)**
11059 Bright Road
Maple City, Michigan 49664
616-228-7159
GLREA holds a renewable energy fair every July, and publishes a quarterly newsletter.

**Midwest Renewable Energy Association (MREA)**
PO Box 249, 116 Cross Street
Amherst, Wisconsin 54406
715-824-5166
MREA promotes renewables and energy efficiency, maintains a network with other energy groups in the Midwest, publishes a newsletter, and sponsors an annual renewable energy fair.

**Northeast Sustainable Energy Association (NESEA)**
23 Ames Street
Greenfield, Massachusetts 01340
413-774-6051 fax: 413-774-6053
NESEA sponsors conferences and workshops on such topics as sustainable building practices, electric vehicles, and renewable energy. It also organizes the yearly Tour de Sol for solar and electric vehicles, and publishes a magazine, the *Northeast Sun*.

**Northwest Conservation Act Coalition (NCAC)**
217 Pine Street, #1020
Seattle, Washington 98101-1520
206-621-0094
NCAC is a regionwide alliance of conservation and consumer advocates, utilities, and individuals working to promote sustainable energy in the Pacific Northwest through research, education, and advocacy.

**Solar Energy Expo and Rally (SEER)**
239 S. Main Street
Willits, California 95490
707-459-1256
SEER sponsors a three-day fair about renewable energy every August. It claims to have the "largest collection of solar and electric vehicles ever assembled."

## STATE AND LOCAL

There are too many organizations to list here, but finding them is just a matter of contacting the right people. In addition to independent groups, there are state and local chapters of national organizations. For example, the American Solar Energy Association (ASES) has twelve chapters. Calling the state energy office, the National Appropriate Technology Assistance Service, national public interest groups, and state groups that you already know of, will likely yield the important players. Printed guides can also help, so see the directories of organizations below.

The local area is also home to some special resources not available on a broader scale. For instance, local entrepreneurs, and people in local planning commissions, municipal utilities, colleges, and universities may be helpful resources, and may also know most of the other local players in the renewable energy arena. Don't forget your local news media—since lots of things go across the desks of reporters, they may be able to help. They may also be instrumental in helping you to publicize your project.

## D. FUNDING RESOURCES

In general, think about who is able to fund your project, and who can potentially gain publicity or other benefits from your project.

## 1. National Funding Resources

### Department of Energy (DOE)

DOE is divided into various organizations, a few of which are able to provide funding for renewable energy projects.

### Department of Housing and Urban Development (HUD)

If your project is a residential building, and the bank thinks that making a loan to you is too risky (because of your innovative renewable energy systems, for instance), it will have you apply for either Title 1 or 203(b) insurance with the Federal Housing Authority (FHA), a division of HUD. FHA will insure the loan so that the bank is not taking a big risk. Title 1 insurance applies to shorter-term loans (seven years or less) of smaller amounts of money ($18,000 or less) to be used for home improvement. These loans have high interest rates, so are generally used in projects which are not too big and which will be paid off quickly. 203(b) insurance covers long-term (thirty-year) loans which are generally used to purchase or refinance a house. Community Development Block Grants are provided to cities for specific projects, as proposed by the city and negotiated with HUD.

### The Energy Foundation

75 Federal Street
San Francisco, California 94107
415-546-7400 fax: 415-546-1794

The Energy Foundation will support policy research and advocacy to identify and overcome regulatory, financial, and institutional barriers to renewable energy development. The Foundation is especially interested in regional initiatives to encourage utility investments in renewable energy through scenario building, regulatory work, and advocacy. The Foundation may also support efforts to catalyze industry-government alliances to develop renewable energy power plants.

### Farmers Home Administration (FmHA)

U.S. Department of Agriculture
Washington, DC 20250
202-447-7967

FmHA provides funding for the construction, acquisition, or rehabilitation of community facilities for rural communities. Communities with fewer than 20,000 people are eligible, as are nonprofit, community-based public interest groups, and Native American tribes. FmHA building codes are very standard with respect to energy conservation; there are no special incentives for renewables and the FmHA generally will require a traditional backup plan for a renewable energy system it is unfamiliar with. FmHA has an office in most states; New England has one regional office.

## 2. State and Local Funding

Various types of organizations may be able to provide funds. You need to become familiar with your area and investigate the potential options. Here are some good places to start:

> Banks
> Businesses and trade organizations
> Community development corporations
> Electric membership cooperatives
> Local charities and foundations
> State and local government
> Universities and colleges
> Utilities

## E. PUBLICATIONS

## 1. Directories of Organizations

*The Activists Almanac: The Concerned Citizen's Guide to the Leading Advocacy Organizations in America,* David Walls. Simon and Schuster, New York, 1993.

*California Environmental Directory: A Guide to Organizations and Resources.* California Institute of Public Affairs, Sacramento, California *Conservation Directory.* National Wildlife Federation, Washington, DC, 1994.
An annual list of governmental and nongovernmental organizations engaged in conservation work at state, national, and international levels.

*Directory of Environmental Groups in New England.* U.S. Environmental Protection Agency, Boston, Massachusetts, 1992. Updated annually.

*Directory of Sustainable Energy Companies.* Sun Day Campaign, Takoma Park, Maryland, 1993.

*Minority Organizations: A National Directory.* Garrett Park Press, Garrett Park, Maryland, 1992.

*National Directory of Sustainable Energy Organizations.* Sun Day Campaign, Takoma Park, Maryland, 1993.

*Rocky Mountain Environmental Directory.* Missoula, Montana. A directory of environmental resources in Montana, Idaho, Utah, Colorado, and Wyoming. Note that this directory is also available on EcoNet as a database.

*The United States and the Global Environment: A Field Guide to American Organizations Concerned with the Environment Issue.* California Institute of Public Affairs, Sacramento, California.

*World Directory of Environmental Organizations.* California Institute of Public Affairs, Sacramento, California.

*World Environmental Directory.* Business Publishers, Inc., Silver Spring, Maryland, 1991. Sixth Edition.

*Your Resource Guide to Environmental Organizations.* John Seredich. Smiling Dolphins Press, Irvine, California, 1991.

## 2. Organizing Resources

*Organizing for Social Change: A Manual for Activists in the 1990s.* Midwest Academy, Seven Locks Press, Cabin John, Maryland, 1991.

*Social and Environmental Change,* Bunyan Bryant. Caddo Gap Press, Davis, California, 1991.

## 3. Media Resources

*All in One Directory.* Gebbie Press, New Paltz, New York, 1994.
Nationwide media directory including papers, radio, TV consumer, business, and trade publications. $80. Updated annually.

*College Media Directory.* Oxbridge Communications, Inc., New York, New York, 1994.

*Editor and Publisher International Yearbook.* Editor and Publisher, New York, New York, 1994. Guide book for advertisers.

*Strategic Media.* Communications Consortium, Washington, DC.
A resource book describing how to use media strategically to promote your cause.

## 4. Books

Akbari, Hashem, et al., *Cooling Our Communities: A Guidebook to Tree Planting and Colored Surfacing.* Washington, DC: Environmental Protection Agency, 1992.

Brower, Michael, *Cool Energy: Renewable Solutions to Environmental Problems.* Cambridge, Massachusetts: MIT Press, 1992.

Gipe, Paul, *Wind Power for Home and Business: Renewable Energy for the 1990s and Beyond.* White River Junction, Vermont: Chelsea Green Publishing Company, 1993.

Gordon, Deborah, *Steering a New Course: Transportation, Energy, and the Environment.* Washington, DC: Island Press, 1991.

Johansson, Thomas, et al., eds., *Renewable Energy: Sources for Fuels and Electricity.* Washington, DC: Island Press, 1993.

Potts, Michael, *The Independent Home: Living Well with Power from the Sun, Wind, and Water.* White River Junction, Vermont: Chelsea Green Publishing Company, 1993.

*St. John, Andrew,* The Sourcebook for Sustainable Design: A Guide to Environmentally Responsible Building Materials and Processes. *Boston: The Boston Society of Architects, 1992.*

Schaeffer, John, ed. and the Real Goods Staff, *Real Goods Solar Living Sourcebook.* White River Junction, Vermont: Chelsea Green Publishing Company, 1994.

Spurling, Walter, ed., *Guide to Resource Efficient Building Elements.* Missoula, Montana: Center for Resourceful Building Technology, 1993.

Strong, Steven J., *The Solar Electric House: Energy for the Environmentally Responsive, Energy Independent Home.* Still River, Massachusetts: Sustainability Press, 1991.

Tumidaj, Les, et al., *Solar Access Design Manual.* Office of Environmental Services, City of San Jose, California, 1992.

Union of Concerned Scientists, *Powering the Midwest: Renewable Electricity for the Economy and the Environment.* Cambridge, Massachusetts: UCS, 1993.

Vale, Brenda and Robert, *Green Architecture: Design for an Energy-Conscious Future.* Boston, Massachusetts: Bulfinch, 1991.

Van der Ryn, Sim, and Peter Calthorpe, *Sustainable Communities: A New Design Synthesis for Cities, Suburbs, and Towns.* San Francisco, California: Sierra Club Books, 1991.

## 5. Periodicals

In addition to the many periodicals published by trade associations, public interest groups, and governmental organizations, see the following:

*Environmental Building News: A Bimonthly Newsletter on Environmentally Sustainable Design and Construction.* West River Communications, Brattleboro, Vermont.

*The Solar Letter.* Alpha Publishing, Silver Spring, Maryland.

*Home Power: The Hands-on Journal of Home-made Power.* Home Power Magazine, Ashland, Oregon.

Nisson, Ned, editor/author, *Energy Design Update: The Monthly Newsletter on Energy Efficient Housing.* Arlington, Massachusetts: Cutter Information Corporation.

## 6. Computer Networks

Recent proliferation and specialization of computer networks allows you to instantly access information and communicate with others regardless of distance. There are many online services which are relevant to renewable energy and other environmental issues. For instance, EcoNet provides environmental news, has numerous conferences, and allows communication between interested people in over seventy countries. (EcoNet, 3228 Sacramento Street, San Francisco, California 94115, 415-923-0900.) For an in-depth look at how to use computer networks, see: *EcoLinking: Everyone's Guide to Online Environmental Information*, by Don Rittner, Berkeley, California: Peachpit Press, 1992.

## F. SUPPLIES AND EQUIPMENT

Harnessing renewable energy often requires a combination of standard equipment and specialized renewable energy equipment. There are many suppliers of specialized goods, some of which may be available in your region. Contact the relevant trade associations to find out who the nearest supplier is. There are also a few mail order companies which may be able to help.

**Real Goods**
966 Mazzoni Street
Ukiah, California 95482
800-762-7325
The largest mail-order supplier of renewable energy products for homes and businesses. Real Goods also carries a wide range of other products which are environmentally friendly.

**Tomorrow's World: Living with the Environment in Mind**
5978 Virginia Beach Boulevard
Norfolk, Virginia 23502
800-461-4739 or 804-461-4739

## Used Equipment

Used equipment can reduce a project's costs dramatically. However, it is important to make sure you are getting what you really need. Zeid Freeman of the Freeman Corporation in Winchester, Kentucky (see page 89) obtained a lot of used hardware for a wood waste boiler system at considerable savings. "Used stuff is a very inefficient market; there's lots out there at good prices but it's very hard to hook into it. You need to have your hands in it or find someone who does." Working with someone who knows the market and who you can trust is a good way to proceed. Depending on your needs, you may be looking for specialized renewable energy equipment or standard hardware. Contacting suppliers in your area and relevant trade associations may help you find what you need.

# Glossary

**alternating current** (AC): electricity that changes voltage periodically, typically sixty times a second (or fifty in Europe). This kind of electricity is easier to move.

**alternatively powered house:** a house which gets its electricity or other power from unconventional sources.

**ambient:** the prevailing temperature, usually outdoors.

**amortize:** calculation of short-term cost over a longer period.

**amp:** measure of flowing electricity.

**amp-hour:** measure of flowing electricity over time.

**audit:** An energy audit seeks energy inefficiencies and prescribes improvements.

**avoided cost:** the amount utilities must pay for independently produced power; in theory, this was to be the whole cost, including capital share to produce peak-demand power, but over the years it has been redefined to be something more like the cost of the fuel the utility avoided burning.

**baseload:** the smallest amount of electricity required to keep utility customers operating at the time of lowest demand; a utility's minimum load.

**berm:** earth mounded in an artificial hill.

BTU: British Thermal Unit, the amount of heat required to raise the temperature of one pound of water 1 degree F; 3,411 BTUs equals one kilowatt-hour.

**buy-back agreement or contract:** an agreement between the utility and a customer that any excess electricity generated by the customer will be bought back for an agreed-upon amount.

**catalytic converter:** a device attached to the exhaust of a vehicle or burner to help complete combustion.

**CFCs:** Chlorinated fluorocarbons, an industrial solvent and material widely used until implicated as a cause of ozone depletion in the atmosphere.

**commercialization:** refers to the process of changing renewable energy use from a novelty to standard practice.

**compact fluorescent (CF):** a modern form of lightbulb using an integral ballast.

**compost:** the process by which organic materials break down, or the materials in the process of being broken down.

**controls:** switches, valves, over-current production (fuses and circuit breakers), and meters that enable us to manage energy systems.

**conversion:** changing energy from one form to another, for example, wind to electricity.

**current:** A predominant energy flow or flow of particles, as in a stream or an electric wire.

**deciduous:** opposite of evergreen, losing leaves in winter.

**degree-days:** the sum, taken over an average year, of the lowest (for heating) or highest (for cooling) ambient daily temperatures.

**Demand Side Management:** conservation or energy efficiency programs initiated by gas and electric utility companies; demand side refers to customer use, while supply side refers to generation sources.

**density restrictions:** imposed by planners, the number of residences permitted on a plot of a given size.

**diesel:** a fuel which powers simple internal combustion engines; it is cheaper than gasoline, and burns dirtier.

**direct current (DC):** the complement of ac, or alternating current, presents one unvarying voltage to a load.

**distribution:** The process and equipment associated with moving energy from where it is delivered to where it is needed; in simple terms, the wires and pipes in the walls.

**DSM:** *See* demand-side management

**Earthship:** a rammed-earth structure based on tires filled with tamped earth; the term was coined by Michael Reynolds.

**efficiency:** A narrow mathematical concept describing the proportion of a resource that can actually be converted into useful project or work; for example, sunlight falling on a PV module contains a given emount of energy, but the module can only convert a percentage of it into electricity.

**electric vehicle (EV):** an automobile powered by an electric motor connected to batteries and/or photovoltaic panels; a non-polluting replacement for fossil-fuel–driven cars.

**electronic ballasts:** an improvement over core/coil ballasts, used to drive compact fluorescent lamps; contains no radioactivity.

**embodied:** of energy, meaning literally the amount of energy required to produce an object in its present form. An inflated balloon's embodied energy includes the energy required to manufacture and blow it up.

**energy:** strictly construed, the potential to do work, that is, to move mass.

**energy efficient mortgages:** This mechanism works two ways: 1) borrowers can amortize the sometimes substantial costs of energy-efficiency improvements by including these costs in the mortgage, recognizing that the utility savings created by the improvements can then be committed to the slightly higher mortgage payments; and 2) lenders can consider the reduced energy costs of an already efficient home when calculating a buyer's debt-to-income ratios, making it easier to qualify for the mortgage.

**energy independence:** an entity or locale enjoys this when it is able to produce any energy it requires without active importation of a source.

**EV:** *see* electric vehicle

**excite the coils:** a generator needs a permanent magnet or an electromagnet to induce a flow of electrons; in a grid-connected generator, energizing the primary electromagnet with power from the grid insures that, if the grid shuts down, the generator stops producing power.

**externalities:** considerations, often subtle or remote, which should be accounted for when evaluating a process or product, but usually are not; for example, externalities for a power plant may include downwind particulate fallout and acid rain, damage to life-forms in the cooling water intake and effluent streams, and many other factors.

**feeder:** *see* subtransmission

**fossil hydrocarbons:** plant and animal material which, having been heated and compressed by overburden (soil, water, or rock) over a geologic period of time (millions of year) has turned to coal, oil, or gas.

**frequency:** of a wave, the number of peaks in a period; for example, alternating current presents sixty peaks per second, so its frequency is sixty hertz. (Hertz is the standard unit for frequency when the period in question is one second.)

**gasifier:** a heating device which burns so hotly that the fuel sublimes directly from its solid to its gaseous state, and burns very cleanly.

**generator:** any device which produces electricity.

**graywater:** all other household effluents besides black water (toilet water); graywater may be reused with much less processing than black water.

**grid:** a utility term for the network of transmission lines that distribute electricity from a variety of sources across a large area.

**grid-connected system:** a house, office, or other electrical system that can draw its energy from the grid; although usually grid-power-consumers, grid-connected systems can provide power to the grid.

**gut rehab:** a construction practice where the interior of an already existing building is torn out and everything is rebuilt.

**head:** the distance water falls from intake to generator in a hydroelectric system.

**headrace:** in a hydroelectric system, the headrace leads water from the forebay to the penstock intake.

**heat exchanger:** device that passes heat from one substance to another; in a solar hot water heater, for example, the heat exchanger takes heat harvested by a fluid circulating through the solar panel and transfers it to domestic hot water.

**high-tech glass:** window constructions made of two sheets of glass, sometimes treated with a metallic deposition, sealed together hermetically, with the cavity filled by an inert gas and, often, a further plastic membrane. High-tech glass can have an R-value as high as 10.

**house current:** in the United States, 117 volts root mean square of alternating current, plus or minus 7 volts; nominally 110-volt power: what comes out of most wall outlets.

**HUD:** Federal Department of Housing and Urban Development.

**HVAC:** Heating, Ventilation, and Air Conditioning: space conditioning.

**hydro turbine:** a device which converts a stream of water into rotational energy.

**incandescent bulb:** a light source that produces light by heating a filament until it emits photons, which is quite an energy intensive task.

**incident solar radiation (or insolation):** the amount of sunlight falling on a place.

**infiltration:** air, at ambient temperature, blowing through cracks and holes in a house wall and spoiling the space conditioning.

**infrastructure:** a buzz word for the underpinnings of civilization—roads, water mains, power and phone lines, fire suppression, ambulance, education, and governmental services.

**insolation:** a word coined from incident solar radiation; the amount of sunlight falling on a place.

**insulation:** a material which keeps energy from crossing from one place to another; on electrical wire, it is the plastic or rubber that covers the conductor. In a building, insulation makes the walls, floor, and roof more resistant to the outside (ambient) temperature.

**Integrated Resource Planning:** an effort by the utility industry to consider all resources and requirements in order to produce electricity as efficiently as possible.

**internal combustion engines:** gasoline engines, typically in automobiles, small stand-alone devices like chain saws and lawnmowers, and generators.

**inverter:** the electrical device that changes direct current into alternating current.

**IRP:** *see* Integrated Resource Planning

**life cycle analysis:** an economic accounting approach that measures the total cost of operating a renewable energy system, factoring in the cost of the equipment and interest on any loans plus the annual cost of fuel, labor and maintenance, all in relationship to the system's life expectancy. This is a more realistic way to look at the cost and savings of an energy system over time, rather than the traditional method which merely considers how long it will take to save enough money to pay back the purchase cost.

**line extensions:** what the power company does to bring their power lines to the consumer.

**low-emissivity (low-E):** applied to high-tech windows, meaning that infrared or heat energy will not pass back out through the glass.

**micro-hydro:** small hydro (falling water) generation.

**modules:** the manufactured panels of photovoltaic cells. A module typically houses thirty-six cells in an aluminum frame covered with a glass or acrylic cover, organizes their wiring, and provides a junction box for connection between itself, other modules in the array, and the system.

**net metering:** a desirable form of buy-back agreement in which the line-tied house's electric meter turns in the utility's favor when grid power is being drawn, and in the system owner's favor when the house generation exceeds its needs and electricity is flowing into the grid. At the end of the payment period, when the meter is read, the system owner pays (or is paid by) the utility depending on the net metering.

**niche markets:** those specialized and relatively small markets where renewable energy applications are currently cost-effective and which, if fully exploited, would greatly advance renewable energy commercialization. Niche markets help increase product demand, thus helping decrease costs and improving product reliability and visibility.

**off-peak energy:** electricity available during the base-load period, which is usually cheaper. Utilities often must keep generators turning, and are eager to find users during these periods, and so sell off-peak energy for less.

**off-peak kilowatt:** a kilowatt-hour of off-peak energy.

**off-the-grid:** not connected to the powerlines: energy self-sufficient.

**ohm:** the basic unit of electrical resistance; I=RV, or Current equals Resistance times Voltage.

**particulates:** particles that are so small that they persist in suspension in air or water.

**passively heated:** a shelter which has its space heated by the sun without using any other energy.

**peak demand:** from the customer's perspective, peak demand is the hottest, muggiest afternoon or the coldest, windiest night of the year when air conditioning or electric heat is needed the most. From the utility's perspective, peak demand is that hour of the year when the combined requirements of all their customers reaches a maximum. for example, many utilities are summer peaking, and summer peaks occur in the mid-to-late afternoon on a weekday after several days of hot weather.

**peak kilowatt:** a kilowatt hour of electricity take during peak demand, usually the most expensive electricity money can buy.

**peak load:** *see* peak demand, above.

**photovoltaics (PVs):** Modules which utilize the photovoltaic effect to generate useable amounts of electricity.

**power:** kinetic, or moving energy, actually performing work.

**power conditioning equipment:** electrical devices that change electrical forms (an inverter is an example) or assure that the electricity is of the correct form and reliability for the equipment for which it provides; a surge protector is another example.

**PUC:** Public Utilities Commission; many states call it something else, but this is the agency responsible for regulating utility rates and practices.

**PURPA:** this 1978 legislation, the Public Utility Regulatory Policy Act, requires utilities to purchase power from anyone at the utility's avoided cost.

**PVs:** photovoltaic modules.

**R-value:** resistance value, usually of materials used for insulating structures. Fiberglass insulation three inches thick has an R-value of 13.

**renewable energy:** an energy source that renews itself without effort; fossil fuels, once consumed, are gone forever, while solar energy is renewable in that the sun we harvest today has no effect on the sun we can harvest tomorrow.

**renewables:** shorthand term for renewable energy or materials sources.

**resistance:** the ability of a substance to resist electrical flow; in electricity, resistance is measured in ohms.

**retrofit:** to install new equipment into a structure which was not prepared for it; for example, we may retrofit a lamp with a compact fluorescent bulb.

**semiconductor:** the chief ingredient in a photovoltaic cell, a normal insulating substance which conducts electricity under certain circumstances.

**set-back thermostat:** combines a clock and a thermostat so that a zone (like a bedroom) may be kept comfortable only when in use.

**solar access:** refers to the various methods for maximizing use of the sun's heating rays, including orienting windows to the south and ensuring that the sun's rays are not blocked from solar collector areas by trees, buildings, etc.

**solar fraction:** the fraction of electricity which may be reasonably harvested from sun falling on a site; the solar fraction will be less in a foggy or cloudy site, or one with a narrower solar aperture, than in an open, sunny site.

**solar hot water heating:** direct or indirect use of heat taken from the sun to heat domestic hot water

**solar oven:** simply a box with a glass front and, optionally, reflectors and reflector coated walls, which heats up in the sun sufficiently to cook food.

**solar panels:** any kind of flat device placed in the sun to harvest solar energy.

**subtransmission:** secondary or feeder elements of the grid that distribute electricity from the major transmission lines to local communities.

**sustainable:** material or energy sources which, if managed carefully, will provide at current levels indefinitely. A theoretical example: redwood would be sustainable if it were harvested sparingly and if every tree taken were replaced with another redwood.

**tailrace:** in a hydroelectric system, the tailrace leads water from the turbine back to the stream.

**therm:** a quantity of natural gas, 100 cubic feet, roughly 100,000 BTUs of potential heat.

**thermal mass:** solid, usually masonry volumes inside a structure which absorb heat, then radiate it slowly when the surrounding air falls below their temperature.

**time-of-day rates:** electric rates that distinguish between electricity used during different times of the day; typically, more is charged for peak-demand times, and less for baseload times.

**transformers:** a simple electrical device that changes the voltage of alternating current; most transformers are inductive, which means they set up a field around themselves, which is a costly thing to do.

**turbine:** a vaned wheel over which a rapidly moving liquid or gas is passed, causing the wheel to spin; a device for converting flow to rotational energy.

**twelve-volt (12-volt):** a kind of direct current electricity, most commonly found in cars, but standard in independently powered homes.

**union:** a plumbing part which allows disassembly without destruction by connecting two pipes together mechanically rather than with solder or glue.

**volt:** measure of electrical potential. 110-volt house electricity has more potential to do work than an equal flow of 12-volt electricity.

**watt:** measure of power (or work) equivalent to slightly less than one thousandth of a horse power, just over three thousandths of a BTU.

# Index

*Renewables Are Ready: People Creating Renewable Energy Solutions* was designed by Jill Shaffer, and set in Sabon, ITC Garamond, and Formata. The text pages were printed with vegetable-based ink on 60# Thor Offset White in Regular finish, by Glatfelter; this is a recycled stock with 85% reclaimed fiber and a 15% post-consumer-waste content. The cover stock is 10-point C1S by Federal. All of the papers in this book are acid-free. The book was printed by Braun-Brumfield, Inc.